Construction Contracts, Questions and Answers

Construction law is sometimes simple and at other times complex. Professionals need answers that are pithy and straightforward but also legally rigorous.

A number of questions come up time and again. They range in content from extensions of time, liquidated damages and loss and/or expense to issues of practical completion, defects, valuation, certificates and payment, architects' instructions, adjudication and fees. Some questions are relatively simple, such as 'what date should be put on a contract?', but there are implications to even the simplest questions. More elaborate ones might involve whether the contractor is entitled to take possession of a section of the work even though it is the contractor's fault that possession is not practicable.

David Chappell is well known as an architect, lecturer and writer on construction law. He has served for many years as Specialist Adviser to the Royal Institute of British Architects, and this experience has given rise to this book. All the questions set out here are real ones; and although most were originally asked by architects, their answers will be of wide interest to QSs, project managers, contractors, employers and others engaged in construction.

David Chappell is director of David Chappell Consultancy Limited and sometime Visiting Professor of Practice Management and Law at the University of Central England in Birmingham. He has written many articles and books for the construction industry, including Spon's *Understanding JCT Standard Building Contracts*, now in its 7th edition. He frequently acts as ⸻

Also available from Taylor & Francis

Understanding JCT Standard Building Contracts
7th edition
D. Chappell

Construction Contracts
Law and management
4th edition
J. Murdoch and W. Hughes

Development and the Law
A guide for construction and property professionals
G. Bruce-Radcliffe

Dictionary of Property and Construction Law
R. Hardy-Pickering

The Presentation and Settlement of Contractors' Claims
2nd edition
M. Hackett and G. Trickey

For price availability and ordering visit our website
www.sponpress.com
Alternatively our books are available from all good bookshops.

Construction Contracts, Questions and Answers

David Chappell

Taylor & Francis
Taylor & Francis Group

LONDON AND NEW YORK

First published 2006
by Taylor & Francis
2 Park Square, Milton Park, Abingdon, Oxon OX14 4RN

Simultaneously published in the USA and Canada
by Taylor & Francis
270 Madison Avenue, New York, NY 10016, USA

*Taylor & Francis is an imprint of the Taylor & Francis Group,
an informa business*

© 2006 David Chappell

Transferred to Digital Printing 2008

Typeset in Sabon by
GreenGate Publishing Services, Tonbridge, Kent
Printed and bound in Great Britain by
Cpod, Trowbridge, Wiltshire

British Library Cataloguing in Publication Data
A catalogue record for this book is available from the British Library

Library of Congress Cataloging in Publication Data
A catalog record has been requested

ISBN10: 0-415-37597-5 (pbk)
ISBN13: 978-0-415-37597-9 (pbk)

Contents

Preface to the First Edition

The Royal Institute of British Architects' Information Line was set up on 1 May 1995. The idea was that RIBA members could ring in and be directed to a specialist adviser who would give ten or fifteen minutes of free, liability-free advice to point the architect in (hopefully) the right direction. I have been a specialist adviser to the RIBA since the inception of the service, answering thousands of questions posed by architects. In my career as a consultant I have also dealt with a multitude of problems from contractors, sub-contractors and building owners.

This book sets out 125 questions I have received and includes some of the more common ones, together with a few unusual ones and frequent misconceptions. Sometimes, there is effectively a ready-made answer, either in the relevant contract or in the judgment of a court. Other questions have no ready answer and in such cases, I have offered a view.

Although, obviously, some of the questions were concerned with earlier forms of contract, they have all been updated to refer to the latest 2005 series of JCT contracts, i.e. SBC, IC, ICD, MW, MWD and DB. Questions have been included on related topics such as architects' fees, design and disputes. Legal language has been avoided. Reference has been made to legal cases so that anyone interested may do some further reading and a full table of cases is included at the back of the book. The contractor is assumed to be a corporate body and has therefore been referred to as 'it' throughout.

This book should be useful to architects, project managers, quantity surveyors, contractors and those building owners who are anxious to understand more about the workings of building contracts.

My thanks to Michael Dunn BSc(Hons) LLB LLM FRICS FCIArb and Michael Cowlin LLB(Hons) DipArb DipOSH FCIArb Barrister who helped me in various ways including the location of some cases.

David Chappell
Wakefield
January 2006

1 Pre-contract issues

1 Can the lowest tenderer legally do anything if its tender is not accepted?

Most invitations to tender contain a proviso that the employer does not guarantee to accept the lowest or any tender. It has long been thought that this allowed the employer considerable freedom to award the contract as desired. To some extent that is correct, but it is not the whole story and employers should take care when tenders are invited that they do not leave themselves open to actions for breach of contract.

In *Blackpool & Fylde Aero Club v Blackpool Borough Council*,[1] the Court of Appeal set out the position when tenders are invited. The position is this. The contractor, by submitting a tender, enters into what can best be described as a little contract with the employer on the basis that, in return for the contractor submitting a tender, the employer will deal with the tender in accordance with the procedure set out in the invitation. At the very least, the contractor is entitled to expect that each properly submitted tender will receive proper consideration. An employer who does not properly consider each tender will be in breach of contract.

Unfortunately, it is not uncommon to find that an employer wishes to see all submitted tenders, even a tender that has been submitted after the closing date and time specified in the invitation. Whatever the architect or quantity surveyor might say, the employer may insist on seeing the tender. On discovering, perhaps, that the late tender is lower than the others, the employer will almost certainly wish to accept it; after all, that is the commercial thing to do.

1 [1990] 3 ALL ER 25

If this tender is accepted, the employer will be in breach of contract, because the others were invited to tender on the basis that only tenders submitted before the closing date would be considered. The submission of tenders created a succession of contracts, each of which included that term. A contractor who learns that the employer acted in breach of contract would be entitled to claim damages. Such damages would certainly embrace all the contractor's costs in preparing the tender. If all the tenderers discovered the breach (and if one did, it is reasonably safe to assume that they all would), the total damages could be considerable.

There may be other stipulations in the invitation, for example about the course of action to be taken if an error is found in the pricing document. Failure to observe these stipulations will also make the employer liable to any tenderers disadvantaged as a result. Quite apart from legal liability, an employer who indulges in this kind of practice will soon find that no contractor is willing to submit a tender on future projects.

However, if the employer strictly observes the rules set out in the invitation, neither the lowest nor any other tenderer has grounds for legal action if a tenderer other than the lowest, or even no tenderer at all, is accepted.

Architects and quantity surveyors who find themselves having to deal with clients who show complete disregard for the tender process must seriously consider whether they can continue to act for such clients. Construction professionals must conduct themselves with complete integrity; this should be an end in itself. In addition, professionals who become associated with doubtful tendering practices will get an unenviable reputation among contractors with whom they will have to work in the future.

2   Is the architect responsible if the tender comes in over budget?

Many architects are concerned about what they should do if tenders come in above the client's budget price. The case of *Stephen Donald Architects Ltd v Christopher King*[2] is instructive. Mr King engaged the architects to deal with the redevelopment of a building into a studio and several flats. Mr King subsequently terminated the architects' services because the cost of the work would exceed the budget. Mr King engaged other architects. As part of the counterclaim, Mr King alleged that the flats were negligently designed, being over-elaborate in the use of materials and space and too costly to proceed. Moreover, it was alleged, the architects did not take proper steps to bring the design within the budget.

The court had some sensible things to say about those allegations. The flats were in an area of local authority housing and it was reasonable for the architects to design some luxurious features to attract buyers into an area that might not seem particularly attractive. Although other architects might have dealt with the brief in a different manner, it could not be said that the architects in this case had produced a design such as no reasonably competent architect would have done.

Once it became apparent that the cost would exceed the budget, the court's view was that the architects should have met with the preferred contractor to try to negotiate a lower price. The architects actually did so and succeeded in making a reduction of some £470,000, producing an achievable construction cost of some £1.3 million. At that stage, the court considered that the architects quite properly waited the outcome of the second tender stage. When that was disappointing, they should have approached the other contractors to see if a better price could be obtained. Although the architects started to do that promptly, Mr King engaged other architects at that point. Therefore, there was no substance in the allegations of negligence against the architects.

So many clients become upset when the tender comes in over budget that it is important for architects to make sure that they work within the budget by having regard to the quantity surveyor's estimates of cost. Where clients add to the brief, architects should confirm the additions in writing and advise clients of the fact that costs are increasing. Invariably at tender stage, clients remember only the original cost estimate: they do not remember asking for changes and richer materials.

Obviously, an architect will be responsible if the tenders show a large increase over the client's budget and there has been no interference and no increase in the client's requirements. Obviously, everything must be taken into account. There may be so much work about that it is difficult to get any contractor to tender and all the tenders are considerably above what might normally be expected. On the other hand, an architect should be sufficiently informed, through the quantity surveyor preferably, so that the client can be warned in advance of the likely level of tenders.

3   The contractor's tender states that it is open for acceptance for six weeks from the date of tender, but the contractor withdraws it after three weeks citing a suddenly increased workload. Is the contractor liable to the employer for the additional costs of a replacement contractor?

The answer to this question is to be found in the law of contract.

When a contractor submits a tender, it is an offer to carry out the required work for a certain sum. The employer is free to accept the offer, reject it or to attempt to negotiate. Until the offer is accepted there is no contract. The law is that an offer can be withdrawn at any time before it is accepted and there are some rather awkward rules regarding acceptance by post. Therefore, in normal circumstances the contractor can withdraw the tender before it is accepted and, strictly, no reason need be given. The contractor has no liability for any additional costs suffered by the employer.

The position is different if the employer pays the contractor to keep the tender open. Tenders often state that 'in consideration of a payment of £1 (receipt of which is hereby acknowledged) the contractor agrees to keep the tender open for acceptance for a period of x weeks from the date hereof'. The effect of that is to create a little contract between employer and contractor whereby the consideration is the employer's payment of £1 and the contractor keeping the tender open. Effectively, the employer has bought an option for a few weeks to decide whether or not to accept the contractor's tender. A sum of £1 may not seem much, but the law does not require that adequate consideration is given. It is sufficient if the consideration has some value. In this case, a contractor who withdraws the tender after three weeks would be in breach of the little contract and the employer would probably be able to bring an action for damages. The damages would be likely to be the additional costs incurred by the employer in engaging another contractor for the work.

Many employers are not aware that the contractor's offer is also brought to an end if the employer rejects it. An employer cannot reject the offer and subsequently, after undergoing a change of mind, decide to accept it after all. In that situation, what the employer may believe to be an acceptance is actually an offer on the part of the employer to form a contract on the basis of the contractor's original offer. No contract is formed until the employer's offer is unequivocally accepted by the contractor.

4 The employer is in a hurry to start work. Is there a problem in the issue of a letter of intent?

It is difficult to think of any other root cause responsible for more difficulties and disputes in construction contracts than the employer being in a hurry. The employer's professional advisers should firmly disabuse the employer of the notion that construction can be put underway (successfully) without proper preparation.

Commonly, the architect will try to overcome the problems of a premature start by issuing a letter of intent. Usually, the contractor will have submitted a tender and it will be referable to a standard form of contract, specification and possibly bills of quantities. The issue of the letter may be due simply to the fact that the employer cannot wait the additional few days necessary for the preparation and execution of a formal contract. If that is the only problem, it can be overcome by a simple letter of acceptance of the contractor's tender rather than a letter of intent. More often, there is something more substantial preventing the issue of an acceptance letter. It may be a delay in obtaining funding for the whole project or perhaps the tender was too high and reduction negotiations are in progress.

The idea of a letter of intent is straightforward. It tells the contractor that the employer is not in a position to enter into a contract for the work, but that work can begin and be carried out in accordance with the drawings and specification and if the employer has to stop the work the contractor will be paid for what has been carried out.

There are several problems associated with so-called letters of intent:

- If it is not carefully drafted, on the contractor commencing work it may create a binding contract for the whole of the work on the basis of the contractor's tender. Simply by putting the words 'Letter of Intent' in the letter heading does not produce a letter of intent.

- If it is properly drafted, either employer or contractor can simply bring the arrangement to an end without notice. This can cause tremendous problems if work has been proceeding under a letter of intent until the work is almost complete. For the contractor to walk away at that stage is very expensive for the employer.
- Although work done under a letter of intent is commonly valued and paid on the same basis as the contract that was envisaged, there is no golden rule about it. Indeed, the contractor is normally entitled to a *quantum meruit,* which may be valued in several ways.
- Sometimes the letter of intent is so carefully drafted that both parties are bound by it until the work is completed although that was almost certainly not the intention.

Letters of intent are sometimes referred to as unilateral contracts, or 'if ' contracts. That is a contract formed on condition: 'If you build this wall, I will pay you £100.' If the wall is built, I am obliged to pay the £100, but there is no contract until the condition is fulfilled.

A letter of intent may constitute a continuing offer: 'If you start this work, we will pay you appropriate remuneration.' Again, there is no obligation on the other party to do the work and, if it is done, there are no express or implied warranties as to its quality.[3]

Hall & Tawse South Ltd v Ivory Gate Ltd[4] is a good example of the problems that can arise when projects are commenced using what one or possibly both parties thinks of as a letter of intent.

Ivory Gate engaged Hall & Tawse to carry out refurbishment and redevelopment works. It was intended that the contract should be in JCT 80 form with Contractor's Designed Portion Supplement and heavily amended clause 19. The tender provided for two stages. In view of the need to start work on site as soon as possible, Ivory Gate sent a letter of intent to Hall & Tawse agreeing to pay 'all reasonable costs properly incurred ... as the result of acting upon this letter up to the date you are notified that you will not be appointed'. The letter proceeded to explain the work required and evinced an intention to enter into a contract in a specified sum.

---

3 *British Steel Corporation v Cleveland Bridge Company* [1984] 1 All ER 504
4 (1997) 62 Con LR 117

Agreement was not quickly reached and Ivory Gate sent a further letter of intent. What was envisaged was that work would commence, contract details would be finalised and the signed building contract would be held in escrow (a situation where the effectiveness of the contract is subject to a condition being fulfilled). Unfortunately the contract documents were never completed. The terms of the second letter of intent were quite detailed, expressing the intention to enter into a formal contract, but pending that time instructing the building contract works to commence, materials to be ordered and Hall & Tawse to act on instructions issued under the terms of the building contract. Previous letters of intent were superseded and, if the works did not proceed, Hall & Tawse were to be paid all reasonable costs together with a fair allowance for overheads and profits. Two copies of the letter were provided and Hall & Tawse were to sign one copy and return it. They neither signed nor returned the copy.

The project took about nine months longer than was planned. Liability was disputed and, therefore, the money due to Hall & Tawse was also disputed. At the time of the trial, the work was nearing completion. The judge referred to the second letter of intent as a provisional contract and said that it had been made when Hall & Tawse accepted the offer contained in it by starting work on site. It enabled the contract administrator to issue any instructions provided the instructions would be valid under the terms of the contract. The judge held that no other contract had come into existence to supplant the provisional contract and the method of determining the amounts due to Hall & Tawse was to refer to the bills of quantities that were to have formed part of the contract. The machinery for valuing the work was to be found in the JCT 80 contract. Under the provisional contract, Hall & Tawse were not entitled to stop work at any time, as would have been the case under a normal letter of intent.

There were two such letters issued in this instance: one was a true letter of intent; the other was actually a contract that determined the rights and duties of the parties. Although it was intended to be provisional until a permanent contract could be executed, the absence of a subsequent permanent contract turned the provisional contract into a permanent contract. A straightforward letter of intent would have entitled the contractor to walk off site at any time and, crucially, it would have entitled the contractor to remuneration on a fair commercial rate basis, which might have exceeded the contract rates.

Manchester Cabins Ltd v Metropolitan Borough of Bury[5] concerned a letter of intent that was not a contract. Tenders were
invited on the basis of the JCT Standard Form of Building Contract
With Contractor's Design. Although Manchester Cabins submitted
a tender, it did not include the contractor's proposals. Much nego-
tiation took place and Manchester Cabins produced some
drawings. Bury sent a fax which stated: 'I am pleased to inform you
that the Council has accepted your tender for the above in the sum
of £41,034.24, subject to the satisfactory execution of the contract
documents which will be forwarded to you in due course.'

Eventually, it was confirmed that the letter was indeed authority
to commence the necessary preliminary works 'subject to the satisfactory execution of the contract documents ...'. Surprisingly, later
on the same day that the confirmation was sent, Bury wrote to
Manchester Cabins suspending work, later stating the Council's
intention to withdraw from the contract. The court held that there
was no concluded contract, because the phrase: 'subject to the satisfactory execution of the contract documents' was included.
Although the phrase did not always prevent a contract from coming into effect, in view of the surrounding circumstances it was
clear that there was no agreement in this instance.

Starting work on the basis of a letter of intent or terms incorporated by reference are, therefore, clearly recipes for litigation. It is
far better for an employer and a contractor to enter at an early
stage into a formal agreement in the current JCT or other form
accepted by both parties.

5   If a letter of intent is issued with a limit of £20,000, is the employer obliged to pay a higher sum after allowing a contractor to exceed the limit?

Letters of intent commonly stipulate a maximum figure that the
employer is prepared to pay. That is perfectly understandable. The
employer needs to know the extent of any financial liability. Thus,
the proposed contract sum may be several million pounds, but,
pending final agreement on contract terms and other matters, the
employer might issue a letter of intent indicating that the contractor

5 (1997) unreported

may proceed up to a total of £50,000 or whatever sum is deemed appropriate. The idea is that, before the contractor completes work to that value, either the contract is agreed and executed or the work is stopped.

The problem is that, once a letter of intent is issued, both parties tend to forget what it says and simply get on with the project as though a contract had been signed. Then something happens that concentrates minds and there is a dispute. *Mowlem plc v Stena Line Ports Ltd*[6] is a case in point. The letter of intent concept was taken rather far by the issue of some 14 such letters during the course of the Works. Fortunately, the parties agreed that each letter superseded the previous one, otherwise the dispute might have been labyrinthine in its complexity. When Mowlem commenced the carrying out of the work described in each letter, a small contract was formed by which Stena agreed to pay Mowlem a reasonable sum. In each case, the maximum amount of each sum was stated in the letter.

The last letter sent by Stena stipulated a maximum amount of £10 million and a date for completion. The Works were not finished by the due date and Mowlem's position was that the work carried out was worth more than £10 million. Mowlem maintained that Stena had allowed it to continue the Works even though it was clear that the cost was exceeding the amount in the letter of intent, therefore, Mowlem ought not to be bound by the amount in the letter, which should not have any effect once the sum was exceeded. Stena contended that its professional advice was that the work done did not exceed £10 million.

The court had no hesitation in concluding that Mowlem was entitled to be paid the reasonable amounts it could substantiate under the terms of the letter of intent but such amounts could not exceed £10 million in total. It would not make commercial sense if an agreement to a maximum sum could be set aside simply because the contractor continued to work after the due date or after the limit had been reached.

From this it is to be concluded that letters of intent, like other contractual documents, mean what they say. Usually, if the contractor is working to a letter of intent that specifies, say, £20,000 as the limit, this figure will be exceeded at the contractor's peril. Of

6 6 October 2004 unreported

course, that is subject to the usual overriding proviso that each set of facts must be considered on its own merits. Where the maximum is low and the eventual sum would be many times that amount, it may be held that a contractor who substantially exceeded the maximum would be entitled to payment on the basis that both parties had clearly agreed to ignore the limit and continue the Works, the failure to issue a revised letter of intent or even a formal contract being an oversight. It is suggested that this would particularly be the case if the contractor had actually received payment above the maximum amount.

6   What date should be put on a building contract?

The date to be put on a building contract is the date the contract is executed. The contract is executed when the last person to sign has signed.

In practice, building contracts are usually prepared by the employer's professional advisers, normally by either the architect or the quantity surveyor. Ideally, it should be prepared by the architect, because the architect is the person who is going to have to administer it.

Clearly, it is always advisable for the contract to be executed before the contractor commences on site. The consequences of failure to enter into a binding contract have been explored in question 4. The question often arises whether a contract that is not executed until halfway through the construction of a project applies only to the part of the project which is constructed after the date of the signature. This sometimes causes the employer to have the contract backdated to a date no later than the date of possession. This is bad practice for several reasons:

- It is stating something that is false – something to be avoided in any circumstances.
- Circumstances may well arise later when it will be important to know the date on which the parties executed the contract. For example, the employer or the contractor may subsequently bring arguments about the date for possession and it may be useful for the other party to be able to show that the contract was signed by both parties many months after possession and that any difficulties could have been raised at that point – indeed, that one or other party could have refused to sign.

• It is unnecessary, because a building contract will be retrospective to cover the whole of its subject matter.[7] Essentially, that is because the parties, even if signing after part of the contract Works have been completed, are signing an agreement to pay for and to carry out and complete the whole of the Works. Therefore, the contract must apply to what has been done as well as what is yet to be done. There may be a sting in the tail here if, for example, the contractor has failed to comply with the requirements of the contract in some of the work already carried out, perhaps because the contractor at that time harboured the idea that the contract would never be signed and, when signing subsequently, forgot the earlier non-compliance.

7 Does the architect have any particular duty to draw the attention of the contractor to onerous terms or amendments in the contract?

If there are onerous or unusual terms or amendments in the contract, the time to bring them to the attention of the contractor is at tender stage so that the terms or amendments in question can be taken into account in the contractor's tender. If the architect waits until after the contract is executed and the contractor has begun or is about to begin work on site, it will be too late.

The position with regard to onerous terms is fairly straightforward. In general, the contractor will be bound by all the terms of the contract that were notified by the employer at tender stage or, at any rate, before the contract was executed. It is usually sufficient if the contractor is notified by means of the bills of quantities or specification. The part referring to the terms applying in each case is called the 'preliminaries' It is here that the contractor is informed of the contract to be used and of any changes to the clauses, for example a change in the period for payment from 14 days to 28 days. It is immaterial whether or not the contractor actually read the terms, so long as the existence of the terms was known.

The point about using standard forms of contract or setting out bills of quantities in accordance with the Standard Method of Measurement 7th Edition, is that contractors know what to expect.

---

7 *Tameside MBC v Barlow Securities Group Services Ltd* [2001] BLR 113; *Atlas Ceiling & Partition Company Ltd v Crowngate Estates* (2002) 18 Const LJ 49

They know what the clauses say and they know what will be included in the bills of quantities and where. If the National Building Specification is used, even the wording of the various paragraphs can be reasonably anticipated.

If it is thought desirable to introduce changes to the standard contracts by amending clauses or even introducing new clauses, it will usually be good notice to the contractor if they are put in the usual places. The exception is if the change or additional clause is particularly onerous. What constitutes 'particularly onerous' will be decided ultimately by an arbitrator or judge, or temporarily by an adjudicator. No rules can be laid down about what constitutes onerous. Questions to be asked might include whether it removes important rights from the contractor or introduces significant duties, or whether it gives the employer significant new rights or removes some normal duties. However, the architect and the quantity surveyor must do their combined best to establish before tender stage whether a clause is onerous. If it is decided that it is onerous, steps must be taken in the documents to give proper notice or, to put it in plain words, to bring it to the attention of the contractor. For example, it must not be buried away in the small print. Not only must it be where the contractor would normally expect to find it, it must also be highlighted in some way. Perhaps it should be placed at the beginning of the document or, in extreme cases, be referred to in the covering letter inviting tenders. Fifty years ago, Lord Denning famously said:

> 'Some clauses which I have seen would need to be printed in red ink on the face of the document with a red hand pointing to it before the notice could be held to be sufficient.'[8]

If generally accepted standard contracts are not used, conditions printed on the back of a letter without any reference to them on the front may be held not to apply.[9] In one instance, a quotation was sent by fax with conditions on the reverse. The reverse of the page was not transmitted and a court held that the reference on the quotation to conditions on the reverse was not sufficient notice.[10] On

8 *J Spurling Ltd v Bradshaw* [1956] 1 WLR 461
9 *White v Blackmore* [1972] 2 QB 651
10 *Poseidon Freight Forwarding Co Ltd v Davies Turner Southern Ltd* [1996] 2 Lloyd's Rep 388

the other hand, it is not necessary that the conditions are set out in full in the document, provided that proper notice of them is given.[11] That is the usual situation when terms are simply referred to in bills of quantities. Obviously, onerous terms cannot be referred to in this way unless the contractor is given plenty of opportunity to inspect the actual terms. However, it is always advisable to set out onerous terms in full in the tender document.

The architect's and probably the quantity surveyor's (or the project manager's if there is one) duty is owed to the employer and not to the contractor. It is part of that duty to ensure that the contractor is aware of all the terms, so that the contract is properly binding on both the parties. If, after the contract has been signed or a tender has been accepted, an onerous clause is discovered by the contractor in the depth of the tender documents where a contractor might not easily notice it, the chances are that it will not apply. It is not the slightest use for the architect or quantity surveyor to draw it to the contractor's attention at that stage; it will not be one of the terms in the contract.

8   Can there be two employers on one contract?

It is presumed that 'two employers' means two entirely separate persons or companies. For example, two friends may jointly buy an old barn and intend to convert it into two dwellings for their families. They may wish to, jointly, engage the architect and, jointly, enter into a contract with the contractor. They may reason, with some justification, that if architect and contractor can be sure of doing both dwellings, there may be financial economies of scale.

The straight answer to the question is 'Yes'. However, there are considerable difficulties involved. The architect will require instructions from the clients and the contractor will require instructions from the architect during the progress of the Works. Therefore, the clients must agree about everything. Who will be responsible for paying? Someone must actually sign the cheques. Will a joint account be set up and will both clients have to sign each cheque? What if the clients disagree? To whom will the architect send fee accounts? Will both clients' names be inserted in the building contract as 'Employer'. What if one of the clients wishes to spend

---

11 *Ocean Chemical Transport Inc v Exnor Craggs (UK) Ltd* [2000] 1 Lloyd's Rep 446

more money than the other? Must there be separate accounts within both architect's engagement and the building contract? What is the position if the two friends have a spectacular falling out?

Most of these questions suggest awful situations, possibly resulting in nightmarish legal proceedings. The architect might have a dispute with only one of the employers, but be obliged to take action against both, because the 'Employer' in the terms of engagement is identified by two names.

An architect or a contractor who agrees to contract on this basis probably has a death wish – certainly an insolvency wish. Although, like a good many other recipes for disaster, it can be done, it is not a good idea. A far better idea, in fact the only sensible idea from the architect's and contractor's points of view, is for the two persons to enter into a contract between themselves which sets out how they will do everything connected with the project. Most importantly, it will say which of the persons will be entered into the terms of engagement and act as client. It will also say that this same person will be entered into the building contract as the 'Employer'. In that way, both architect and contractor will have one point of contact and, to put it bluntly, one person to sue if things go wrong. How the two persons arrange their own liabilities is their affair, but not something that need concern the architect and the contractor.

9   If the employer wishes to act as foreman, can each trade be engaged on an MW contract?

It is becoming more common for employers to wish to contract separately with the various trades engaged in the construction process. During the nineteenth century in Britain it was the norm for employers to engage an architect to design a building, following which the architect, on the employer's behalf, would hire the various trades to carry out the work and a site foreman to control it. Subsequently, main contractors became usual, each one engaging sub-contractors or actually employing the relevant tradesmen directly.

An employer who wishes to engage separate trades and act as site foreman must be sure of having the requisite skills. Although the employer no doubt believes that large sums will be saved, which were added to the bill simply by the existence of a main contractor, it is easy to lose those large sums quite quickly if the progress on site develops serious problems.

The fact is that there is no currently available contract that precisely caters for this situation. There are, of course, a multitude of sub-contracts but, not surprisingly, they are all designed to be used with a main contract and a main contractor is assumed. The Trade Contract for use with the Construction Management Contract is more likely, although it still assumes the existence of a construction manager.

Many employers in this situation look to the minor works contract. Although it is superficially an attractive proposition, there are many pitfalls. The most important point is that it was not written to address the contractual relationship between the employer and each of a number of separate sub-contractors. It assumes that there is but one contractor on the site carrying out the Works. There is an implied term that the contractor has exclusive possession of the site, which would have to be amended to provide that the particular contractor acknowledges that it is simply one amongst several. Otherwise, there will be claims from contractors alleging the others are interfering with the progress of their work. It assumes that instructions will be given by the architect or contract administrator. If the employer really wishes to act as foreman (and has the ability to do so), there must be a clause in the contract to permit the employer to direct each contractor in its work and to require each contractor to comply with such directions. Virtually every clause would require serious amendment. The liquidated damages clause is unlikely to be applicable and certainly there would be problems with the insurance clauses and termination provisions, which cite employer interference as one ground for the contractor to issue a default notice.

In brief, so many amendments would be necessary to the Minor Works Building Contract and so many additional clauses would be required to deal with such matters as attendances and claims for loss and/or expense that it would be easier for the employer in this situation to commission the drafting of a special contract especially for the purpose. The employer would then have to face the possibility of trade contractors refusing to contract on unfamiliar terms.

10 What if no one notices the contractor's serious financial error until the contract is executed?

Tendering is a process conducted at breakneck speed, or so it seems to the contractors concerned. The period allowed for tendering, even if it complies with industry accepted norms, is never really

long enough for all the complicated obtaining of sub-contract tenders and the taking into account of the myriad of clauses in the
specifications, drawings or bills of quantities. It is little wonder that
there are mistakes in tenders. The wonder is that there are not more
mistakes.

If there is a mistake that is noticed before a tender is accepted,
it must be dealt with in accordance with the particular tendering
process adopted. For example, if the quantity surveyor has stipulated that tendering will be in accordance with the CIB Code of
Practice for the Selection of Main Contractors (1997), the contractor's act in submitting a tender will create a contract
whereby the employer, through the quantity surveyor, agrees to
properly consider and apply the Code to any tender properly
submitted.[12] Failure to apply the Code in these circumstances
will entitle the disadvantaged contractors to take action against
the employer for damages. What such damages might be in any
particular circumstance is open to debate, but it might, at the
very least, allow the contractors to claim the cost of tendering.

The situation for the contractor is grim if the employer has properly complied with all the relevant procedures, but the contractor's
serious mistake has not been noticed until after the contract documents have been executed. In those circumstances, the contractor
has no remedy. The situation was considered as long ago as 1927
in *W Higgins Ltd v Northampton Corporation*,[13] when the court
sympathised with the contractor, especially because the mistake
was 'really brought about by the carelessness of some official of
[Northampton] in drawing up the original bill of quantities'.
Unfortunately for the contractor, that carelessness did not alter
the fact that the contractor had put in a price substantially lower
than intended, it had been accepted and the contractor was
bound by that mistake.

The situation is different where the employer realises that the
contractor has made a mistake in the tender, but tries to conclude
a contract embodying the mistake. Such a situation arose in
McMaster University v Wilchar Construction Ltd,[14] where the

---

12 *Blackpool and Fylde Aero Club v Blackpool Borough Council* [1990] 3 All ER
    25
13 [1927] 1 Ch. 128
14 (1971) 22 DLR (3d) 9

contractor inadvertently omitted an important page of the tender. The page included a price fluctuations clause. The contractor soon realised the error and notified the employer which, nonetheless, tried to accept the tender. The court severely criticised the employer and held that the contract was voidable for a fundamental mistake in its formation. The position is probably the same even if the contractor has not given prior notice to the employer if it can be shown that the employer knew of the mistake in any event.

Therefore, contractors must take great care in preparing and submitting tenders. The *McMaster* case is no doubt unusual. In most cases, quantity surveyors or other relevant professionals subject all tenders to detailed scrutiny and pride themselves on finding errors. Nevertheless, contractors should not rely on the expertise of quantity surveyors, because if errors are missed and the contract is executed, the erroneous sums become binding on both parties. It is less likely that an error would be in favour of the contractor, unless it was very subtle, or the tender would not be accepted. But if an erroneous price is accepted, it is binding even if the contractor receives an unexpected windfall.

2 General contractual matters

11 If the employer is in a partnering arrangement with a contractor, does that mean that the SBC does not count?

Partnering is much misunderstood. First, it must be distinguished from 'partnership', which is a legal relationship between two or more persons acting together with a view to profit. Partnership is governed by the Partnership Act of 1890. Partnering is not a legal relationship at all. It is simply the name that some people have given to the way in which a building may be procured. It is not a procurement system itself, because the procurement of the building might be by means of a traditional arrangement or any combination of project management, design and build or management contracting even though partnering is being practised.

Because partnering is very much a recently invented process, it is not susceptible to clear definition. It may comprise one or more of a whole host of constituent parts. For example, the parties may agree an open book policy whereby the contractor agrees to allow the employer's professional advisers access to its books of account so that they can see whether the contractor is making a profit and how much, and the actual cost to the contractor of labour and materials. There is usually a requirement that all parties will act in a spirit of mutual co-operation and trust and will give each other early warning of problems. Sometimes the arrangement includes provision for both parties to share in any savings and sometimes for the employer to shoulder part of a contractor's loss – up to a certain figure. A guaranteed maximum price (GMP) is often a feature. Like many other things, this is usually not as good as it sounds; a GMP is not necessarily the maximum price and is rarely guaranteed. It has been said, with some truth, that partnering is an attempt to go back to the attitudes and values that existed in the construction industry fifty years ago. The problem is that much has changed in fifty years.

Parties who decide to go forward on a partnering basis do so in one of two ways. First, it is possible for them to enter into a partnering contract, that is to say, a contract that contains within its terms various clauses binding the parties to the partnering ethos. Such contracts as the Engineering and Construction Contract (NEC) and the PPC2000 are examples. A difficulty with both these contracts is that they bear little resemblance to the widely used JCT series of contracts or even the less widely used, but excellent, ACA contract. Therefore, these partnering contracts take a great deal of time to assimilate and use properly. They are both very complex and some may say eccentric. But the parties are legally bound by the partnering principles. Another difficulty is that there is little experience of parties being bound by a requirement, for example, to work in a spirit of mutual trust and harmony. If one party believes another is not being friendly, can the injured party bring legal action for breach of contract? It seems unlikely.

Second, parties may achieve partnering by entering into a legally binding contract in the usual way and incorporating anything that they believe should be legally binding, such as price, access to books, sharing savings and so on. At the same time and usually after all parties have spent a day engaged in getting to know one another, they enter into what is commonly called a 'partnering charter'. This is usually signed at the end of the 'getting together' day whereby all parties (including the professionals) enter into non-binding undertakings that they will work together in a co-operative way. The charter is non-binding, because the kinds of things it contains are precisely those that are difficult to enforce and depend upon all parties understanding that working together in this way is a good way of achieving a satisfactory outcome for all concerned.

The question is based upon a complete misunderstanding of partnering. If the parties enter into a contract such as NEC or PPC2000, it will be binding and breaches of the contract will result in one of the parties being able to bring proceedings in adjudication or either arbitration or litigation. If the parties enter into a JCT contract plus a non-binding charter, they will be bound by the JCT contract. However, what the parties have informally agreed to do in the charter may be taken into account when the way in which the parties have carried out their obligations is considered by a judge or arbitrator.[15]

It has been known for some ill-advised parties to attempt to procure the construction of a building without a legally binding contract at all, but simply relying on expressions of goodwill in a non-binding partnership charter. This is somewhat equivalent to an old fashioned shaking of hands, but in today's climate it is the height of folly. In the event of a dispute, there is no written contract, therefore the provisions of Part II of the Housing Grants, Construction and Regeneration Act 1996 will not apply. Whether there is indeed a contract of some sort, perhaps formed by oral exchanges between the parties, will be a matter for the courts. Even if binding, expressions of goodwill are not in themselves sufficient to form the basis of a contract. There must be a known price for the work, or a way of calculating it, and there must be an agreement on the work to be done and the period of time in which it must be done. The courts may be prepared to imply some terms into a contract, but they are not prepared to create a contract where none existed.

The age of partnering (however long or brief it is before the next answer to the construction industry's difficulties arrives) is not a time for throwing away binding contracts. Without a binding contract, neither party is obliged to start, let alone finish, anything.

12   The contractor has no written contract with the employer (A). (A) instructed the contractor to do work and asked it to invoice their 'sister company' (B). The contractor did so and B has not paid despite reminders.

This issue is in two parts that are not necessarily connected.

It is always a mistake not to have a written contract. Oral contracts depend upon either the parties agreeing on what they said or the presence of reliable witnesses. It should set alarm bells ringing when a person asks for work to be done or services performed, yet is unwilling to enter into a proper contract. The fact that there is no written contract may suggest that there is no contract at all.

---

15 *Birse Construction Ltd v St David Ltd* (1999) 70 Con LR 10 CA; [1999] BLR 194

There may be what is sometimes referred to as a *quasi-contract* – more accurately a claim in restitution. In this sort of situation, the claim arises when one party gives instructions to another to carry out work or perform services, knowing that the other party carries out that kind of work or performs those services for a charge. In those circumstances, the law may hold that there is an implied promise to pay.[16]

The second and more important part of the question concerns the identity of the paying party. Whether there is a contract entered into orally by the parties or whether we are dealing with a simple promise to pay, there will be a legal obligation that, in simple terms, amounts to the contractor doing work for (A) and (A) having a legal obligation to pay the contractor in return.

Although the contractor would normally expect to invoice (A) for the work, in this instance (A) has introduced a new element by asking the contractor to invoice (B). Despite the fact that (A) has described (B) as a 'sister company', if both are limited companies they are separate legal entities. Therefore, (B) can quite properly refuse to pay, because it has no obligation to pay the contractor under a contract or even a *quasi contract*. Therefore, the contractor has no grounds for any legal action against (B) to recover the unpaid charges. Unfortunately, legal action against (A) for the unpaid invoices would be equally fruitless, because the invoices were never sent to (A) and (A) has no obligation to pay an invoice addressed and sent to another party.

Sadly, this situation is not at all rare. If the obligation to pay has been properly assigned to (B) with a written agreement to the assignment by all parties the situation is different, but assignment in these circumstances is not the norm. A letter from (A) to the contractor directing that invoices should be sent to (B) will not assist the contractor, unless it unequivocally states that a failure to pay on the part of (B) will be made good by (A).

Usually, the contractor's only remedy is to re-invoice (A) with the whole amount owing, wait the prescribed period and then, if not paid, to take legal action for recovery against (A). Obviously, the contractor cannot recover interest on late payment under the Late Payment of Commercial Debts (Interest) Act 1998, because

16 *Regalian Properties v London Dockland Development Corporation* [1995] 1 All ER 1005

the invoice will be the first he has sent to the correct party. A recurring problem concerns whether or not a binding contract has been entered into in any given situation and, if so, in what terms. That is fundamental to the resolution of most problems in the construction industry. Unless the rights and duties of the parties can be pinpointed with reasonable accuracy, it is impossible to say whether they have or have not complied with their obligations.

Stent Foundation Ltd v Tarmac Construction (Contracts) Ltd[17] is an example of this kind of problem. It is also of interest because it concerned the JCT Management Contract.

At the time the problem arose, the management contractor was Wimpey Construction Ltd, later becoming Tarmac. The employer was a firm called Wiggins Waterside Ltd. Stent was the prospective works contractor. Stent was certainly employed to carry out foundation works – the question was: by whom? The reason why the question came before the court was that Stent had a large claim for the cost of dealing with ground conditions, but Wiggins was in receivership.

Under the JCT Management Contract, there is generally no difficulty that the works contractor is in contract with the management contractor under the terms of the JCT Works Contract. In this instance, however, the position was clouded, because tenders had been invited and a letter of intent sent to Stent by the employer before the appointment of Wimpey as management contractor. Wimpey, although not formally appointed, had been involved in the foundation works contractor tendering process and confirmed to Stent that it would be in contract with Wimpey.

Representatives of Wiggins continued sending instructions to Stent. A key document was probably a letter of instruction to Stent stating that they were to start work on the foundations under the letter of intent, but that Stent must enter into a Works Contract with Wimpey. From then on, Stent and Wimpey acted as though a Works Contract had been concluded between them, but Wimpey stated, quite reasonably, that it could not enter into a Works Contract until the formalities of the Management Contract with Wiggins had been completed. In the meantime, Wimpey continued to process applications for payment from Stent as though the Works Contract was in place. In fact, no Works Contract was ever formally executed, despite Wimpey having executed the Management Contract with Wiggins.

The judge held that a binding contract came into existence between Wimpey and Stent when Wimpey entered into the Management Contract with Wiggins. Wimpey and Stent had been agreed about all the important terms of their contract before then and the contract was dependent only on the execution of the Management Contract.

The judge rejected a suggestion by Stent that Wimpey were estopped (prevented) from denying the existence of a contract just because Wimpey had acted in all ways as though a Works Contract had been executed. He said that Wimpey might have acted in this way in the expectation of a contract being agreed, which would then, of course, be retrospective. That is a very important point to remember. It has long been a construction industry myth that if both parties act as though there were a contract, there really will be a contract. It is a truism to say that every case depends upon its own particular facts; there is no doubt that, in considering whether in any particular instance there is a contract, every facet of the conduct of the parties has to be considered. Acting in accordance with a supposed contract's terms may reinforce the conclusion that there is a binding contract, but it will not usually be conclusive without other evidence.

13   Is there a contract under SBC if everyone acts as though there is?

Many of the problems in the construction industry would be avoided if the parties concerned ensured that there was a proper written contract in place and signed by both parties before work began on site. Anything else is asking for trouble.

Parties to building contracts seem strangely reluctant to settle all the formalities before work begins. Despite all the evidence to the contrary there is a firm belief that once construction starts on site, the formalities of a legally binding contract will not be a problem. Experience shows that this is a misguided view. In answer to queries about the existence of a formally executed contract an architect will often say that the parties did not get around to actually signing the contract, 'but everyone acted as though the contract had been signed'.

In *A Monk Building and Civil Engineering Ltd v Norwich Union Life Insurance Society*,[18] the existence or otherwise of a binding contract was in issue. Essentially, Monk claimed that there was no

18 (1991) CILL 669, upheld by Court of Appeal (1992) 62 BLR 107

contract and that they were, therefore, entitled to be paid on a *quantum meruit* basis: Norwich argued that there was a contract and Monk were entitled only to the contractual sum. It was originally intended that the contract would be executed as a deed before work commenced. Draft contracts were scrutinised by Monk, which made various amendments. Eventually, a letter of intent was issued to Monk on behalf of Norwich. Monk made various amendments to the letter, signed and returned it and made arrangements to start work without prejudice to the unresolved contractual position. At the time of starting work, there were many items still unresolved. The project managers wrote to Monk shortly after work commenced with what was later described as an offer, which it was alleged Monk accepted by continuing work.

Discussion continued during the progress of the Works, but without agreement on several terms. Throughout the project, both Monk and the project managers relied on various contract provisions.

The court held that there was no concluded contract between the parties. The project managers had no actual or ostensible authority to negotiate a contract on behalf of Norwich, but only authority to issue the letter of intent. Importantly, there was no agreement of all necessary terms. The court said that it was irrelevant that Monk had relied on contractual provisions during the progress of the Works.

Although the reliance by both parties on contract terms might, in some circumstances, indicate that both parties had accepted that they were bound by the contract, that is an assumption which can be overturned by other factors. Key factors are where, as in this instance, there is no evidence of acceptance of the contract and there is clear evidence that important terms remain unagreed.

14   Can certificates and formal AIs be issued if the contract is not signed?

This is a question that often troubles architects when work has begun under a so-called letter of intent, a month has gone by and the contractor is yelling for a certificate. Strangely, the architect has probably issued several Architect's Instructions by this time. It is the request for money which, as usual, concentrates the mind.

Obviously, contracts should always be signed by both parties before any work begins on site. The use of letters of intent is not to be encouraged. They lead to a false sense of security. If the parties had contracted on what might be termed a simple contract where the contractor agreed to carry out work for a price and start and completion dates were agreed, the architect would have no power to issue either certificates or instructions. Indeed, an architect in these circumstances would have no power at all because there would be no mention of the architect in the simple contract. Certain clauses would be implied by statute or under the general law, but the presence of an architect, or a quantity surveyor for that matter, would not be one of them.

If work is being carried out under a true letter of intent, a very limited contract would be formed of the 'if' variety: 'If you do some work, I will pay you a reasonable amount of money.' But few, if any, other terms would be implied and certainly the architect would have no rights or obligations under it.

However, the situation may be that the contractor has been invited to tender on the basis of drawings and specification or bills of quantities and these documents may include the clearest details of the contract to be executed, including how all the contract particulars will be completed. If the contractor submits an unqualified tender on that basis and if the employer proceeds to accept the tender without any equivocation, a binding contract will be formed incorporating all the details of the drawings and other documents in the invitation to tender and, most importantly, incorporating the terms of the contract specified in the documents. The architect will then be able to act exactly as if the parties had executed the formal contract documents.

Of course, if the acceptance or the invitation to tender makes reference to acceptance being subject to the execution of formal documents, no contract is in place until that is done. On the other hand, tenders may be submitted with qualifications or letters of acceptance may include conditions that make them counter-offers, but the qualified tender may be unequivocally accepted or the counter-offer may be accepted by the contractor and a binding contract come into existence in that way. The possible permutations are probably endless and great care is required to properly categorise the relationship.

15 Does a note in minutes of a site meeting rank as written agreement for the purposes of the Housing Grants, Construction and Regeneration Act 1996?

From time to time, questions arise about the status of site meeting minutes. A favourite question is whether they constitute a written instruction of the architect as required by the JCT and other standard contracts. The answer is probably that they do, provided that the minutes are drafted by the architect. A more bizarre question is whether minutes are written applications by the contractor in respect of direct loss and/or expense. The answer to that is clearly that they are not, even if they record that the contractor asked for loss and/or expense at the meeting, unless the contractor actually drafted the minutes.

Whether an agreement is in writing is an important question if a contract is to be brought within the Housing Grants, Construction and Regeneration Act 1996. Section 107 of the Act makes clear that the Act applies only to agreements in writing.

At first sight, it appears that section 107 is drafted very broadly so as to include virtually every contract, even if it could only tenuously be said that it was in writing. For example, a contract made orally, but which refers to a contract in writing satisfies the Act. However, it is now established that it is not sufficient if the contract was made by a mixture of exchanges of correspondence, notes and oral agreements: a complete record of the parties' agreement must be in writing.[19] Of course, that presumably means that if the agreement was fairly basic, it is only what was agreed that need be in writing, subject always to the requirement for the essential terms to be agreed. So, for example, the agreement will not be excluded from the operation of the Act if the written agreement makes no reference to an extension of time clause if the parties did not agree such a clause.

There seems to be little doubt that if minutes were prepared of a meeting during which the parties agreed essential terms, and if such terms were all recorded in the minutes, it would constitute a written agreement for the purposes of the Act. If the contractor submitted a written tender and the order to commence work was recorded in the minutes of a meeting, that would also fall within the Act as written evidence of acceptance of a written tender.[20]

16   If the employer sacks the architect under MW and appoints an unqualified surveyor as contract administrator, is the contract still valid?

The identity of the architect is stated in article 3 of the JCT Minor Works Building Contract 2005 (MW). It is similar to its predecessor

---

19   *R T J Consulting Engineers v D M Engineering* [2002] BLR 217
20   *Connex South Eastern Ltd v M J Building Services Group plc* [2004] 95 Con LR 43

(MW 98) in that it states that if the architect ceases to act, the replacement architect will be the person nominated by the employer within 14 days. There is a proviso that the replacement architect must not disregard or overrule any certificate, opinion, decision, approval or instruction given by the former architect except to the extent that the former architect would have been able to do so.

In article 3, reference is to the 'Architect/Contract Administrator' and this way of describing the person named in article 3 is adopted throughout the contract. Footnote [7] of MW explains that if the person named in the contract is entitled to use the name 'Architect' in accordance with the Architects Act 1997, 'Contractor Administrator' should be deleted and is then deemed deleted throughout the rest of the contract. The purpose of providing the alternative name is to protect an unregistered person from being prosecuted under the Act.

It follows, therefore, that, if the person originally named in the contract is an architect, the words 'Contract Administrator' must be deleted or if not deleted will be deemed deleted there and throughout the contract. Clearly, an unregistered person, even if a qualified surveyor, cannot be appointed as architect. Such an appointment would not invalidate the contract, but it would be unlawful and of no effect. Indeed, the person concerned would be liable to prosecution for infringing the Act and may have to pay a substantial fine.

There is another, separate, consideration. When the contractor tendered, it would have been on the basis that the contract would be administered by an architect of known ability. It is a sound argument that, in the case of any replacement, the replacement person must be of the same ability. This would prevent the all-too-prevalent practice whereby an employer sacks the architect and self-appoints.

If the position was reversed and the unqualified surveyor was sacked and replaced with an architect, there should be no difficulty, because the surveyor would have been described throughout the contract as the 'Contract Administrator' and the architect certainly fits into that category.

Under the JCT Standard Building Contract 2005 (SBC), article 3 states the name of the architect and clause 3.5 provides that renomination must take place within 21 days of the architect ceasing to hold the post. The contractor is given express power to object within 7 days for a reason accepted by the employer or considered to be sufficient by an adjudicator, arbitrator or judge unless the employer is a local authority and the former architect was an official of it. The JCT Intermediate Building Contract 2005 (IC) names the architect in article 3 and provides for replacement in clause 3.4.1 in a somewhat shortened version of the SBC clause but to similar effect, save that the employer has only 14 days to nominate.

17   Can a warranty be effective before it is signed?

There are relatively few cases on warranties and *Northern & Shell plc v John Laing Construction Ltd*[21] settles an important point.

Laing entered into a contract for the construction of an office block. Under the main contract, Laing was obliged to give a warranty in stipulated terms to the company leasing the building from the developer. The successor to this company was Northern. Some years after completion of the building, defects were discovered and Northern relied on the warranty to recover damages from Laing, because the warranty provided that Laing had complied with the terms of the main contract. The warranty stated: '5. This deed shall come into effect on the day following the date of practical completion of the building contract.'

The limitation period for a deed is 12 years and Northern had not started legal proceedings until not long after that period had expired. However, there was a complication in that the warranty had not been signed until 5 months after practical completion and Northern argued that the normal rules applied and the cause of action arose when it was signed. If that was the case, the issue of proceedings would be inside the 12-year period.

The Court of Appeal decided in favour of Laing. It ruled that clause 5 meant what it said. It had been open to the parties to amend the clause when the warranty was signed so as to take account of the fact that the warranty was being signed retrospectively. The fact that the parties did not amend clause 5 indicated that they intended the warranty to come into effect on the day after practical completion.

The lesson to be learned is that, when a contract of any kind is being executed by the parties after the project has begun, it pays to carefully review the wording of the contract just in case any part should be amended. This is frequently the case in regard to dates for possession in building contracts often signed months after possession was given late. In this instance, the effect was that the warranty started at a date before it was signed. If clause 5 had not been included, the warranty would not have been effective until it was signed. Obviously, if it had never been signed, it would never have been effective at all, with or without clause 5.

18 SBC: There is a clause in the bills of quantities preliminaries which states that no certificates will be issued until the contractor has supplied a performance bond. Work has been going on site for 6 weeks and there is no performance bond, but the contractor says that the architect must certify.

This is a surprisingly common provision. The contractor is correct. In *Gilbert Ash (Northern) v Modern Engineering (Bristol)*,[22] this kind of action was held to amount to a penalty that was, therefore, unenforceable, because large amounts of money could be withheld for a trivial breach. It can readily be seen that if a performance bond is required in the amount of 10 per cent of the contract sum and if 10 per cent of the contract sum was, say, £85,000, the contractor might well have earned this amount in two or three months. Therefore, to withhold payment beyond that point would be to penalise the contractor unduly. This is something the courts have never condoned. In considering liquidated damages and penalties, the courts have made clear that a greater sum can never be proper recompense for the loss of a lesser sum.[23]

The Court of Appeal took a down-to-earth view in relation to a contract's commercial aspect in a more recent case involving the Millennium Dome.[24] The contract between Koch and Millennium (for the supply and fixing of the roof) contained a clause which said: 'As a condition precedent to any liability or obligation of the client under this Trade Contract, the Trade Contractor shall provide at its own costs a guarantee in the form outlined in Schedule ...'. The contract documents were completed by Koch, but no guarantee and performance bond was completed. Koch confirmed that a guarantee and performance bond would be completed and sent. Koch then heard that the contract might be given to another company and suspended

22 [1973] 3 All ER 195
23 *Kemble v Farren* (1829) 6 Bing 141
24 *Koch Hightex GmbH v New Millennium Experience Company Ltd* (1999) CILL 1595

the execution of the guarantee and bond until the position was clear. Subsequently Koch's employment under the trade contract was terminated. Millennium then argued that it was not obliged to make any payment to Koch, because the condition precedent was not satisfied.

When the matter came before the Court of Appeal, the judge thought that Millennium's contention was misconceived. The purpose of the guarantee and bond was to ensure that Millennium was protected when the works commenced. The judge stated:

> It was suggested on behalf of the Millennium Company, that the purpose is achieved by relieving the client from the obligation to make any payments until the guarantee and the performance bond have actually been provided. But, as it seems to me, the client and the trade contractor cannot have intended that the effect of their agreement should be that the trade contractor should be entitled to carry on works without being paid for some indefinite period until it chose to provide the guarantee and performance bond. Such an arrangement could properly be described, in my view, as commercial nonsense.

In addition, by choosing to terminate Koch's employment under the contract, Millennium made it impossible for the condition precedent to be fulfilled.

This case makes clear that the courts will take a dim view of very onerous conditions in business contracts if they do not make commercial sense.

19   Can a contractor avoid a contract entered into under economic duress?

The straight answer to this question is 'Yes'. The more important question is 'what is economic duress?' because, unless a contractor can recognise it, the question is academic.

In practice, economic duress is not common. The law recognises that there are certain forms of pressure that may be applied to a party which do not amount to a physical threat to a person nor to damage to goods, but which may allow the innocent party to throw off the contract. That kind of pressure will often be applied when a contract already exists in order to obtain better terms.[25]

---

25 *Carillion Construction Ltd v Felix (UK) Ltd* (2000) 74 Con LR 144

A typical scenario was demonstrated in *Atlas Express Ltd v Kafco (Importers and Distributors) Ltd.*[26] Atlas was a well-known parcel carrier. It entered into a contract with Kafco, a basketware importer, to carry parcels at an agreed rate. The contract, which was recorded by telex, was crucial to Kafco, because it had just entered into a contract to supply basketware to a large number of Woolworth stores. After a few weeks, the carriers' representative decided that the agreed rate was not viable for their company. Atlas knew that Kafco was contracted to Woolworth, which would sue and cease trading with the importers if deliveries were not made. Atlas therefore wrote suggesting that the agreement be updated, and refused to carry any further goods until a fresh agreement was signed.

Kafco protested, but it had no real option but to sign the new agreement. However, when invoices arrived, Kafco paid only the sums calculated in accordance with the original agreement. Atlas took legal action to recover the balance. In deciding in favour of Kafco, the court pointed out that no person could insist on a settlement procured by intimidation. Economic duress was recognised in English law and, in this instance, it voided Kafco's consent to the second agreement. In addition, Atlas had not provided any consideration for the revised agreement and consideration was vital for the establishment of a contract or the variation of an existing contract.

A contract is essentially entered into on the basis of the exercise of free will by both parties. It has long been established that physical violence or the threat of it in order to induce a contract will make that contract void or at least voidable at the option of the threatened party. Economic pressure can be as difficult to withstand as physical violence.

Economic duress must not be confused with undue influence, which is founded on a different principle. In cases where there is some kind of confidential relationship, such as between bank manager and client, the court will usually presume undue influence prevented the client from making an independent judgment if there are dealings between the two parties.

It has been held that the following factors (which must be distinguished from the rough and tumble of the pressure of normal commercial bargaining) must be taken into account by a court in determining whether there has been economic duress:

26 [1989] 1 All ER 641

- whether there has been an actual or threatened breach of contract
- whether or not the person allegedly exerting the pressure has acted in good faith
- whether the victim has any realistic practical alternative but to submit to the pressure
- whether the victim protested at the time
- whether the victim affirmed or sought to rely on the existing contract.[27]

Economic duress is the wielding of economic sanctions to induce a contract. There are many instances in the construction industry where contractors or sub-contractors are persuaded to make savings or carry out additional work in circumstances that border on economic duress.

20   Funding has been stopped and three certificates are unpaid. The contractor has suspended obligations. Subsequently, vandalism occurred on site – whose problem is that?

Section 112 of the Housing Grants, Construction and Regeneration Act 1996 provides that if a sum due under a construction contract is not paid in full by the final date for payment and no withholding notice has been served, the person to whom the sum is due has the right to suspend performance of all obligations under the contract.

A contractor engaged on a construction contract that does not have a residential occupier is entitled to enforce that right by giving at least 7 days' written notice to the employer. The right is included in SBC by clause 4.14, in IC by clause 4.11 and in MW by clause 4.7. The principal difference from the Act is that each of the JCT clauses requires the contractor to send a copy of the suspension notice to the architect. If that is not done, the suspension is still valid, but it is carried out under the Act rather than under the contract terms. Where a copy is sent to the architect, the suspension is carried out under the terms of the contract that provide, in the case of SBC and IC, that the contractor will be entitled to an appropriate extension of time and any loss and/or expense caused by the suspension. The Act provides only for extension of the contract period.

27 *DSNB Subsea Ltd v Petroleum Geoservices ASA* [2000] BLR 530

In this question, it seems that the contractor has exercised a remarkable – one could say foolish – degree of forbearance so far as the three unpaid certificates are concerned. It is presumed that the requisite 7 days' written notice has been given; if not, then the situation would be less clear. Failure to pay one certificate is a breach of contract on the part of the employer, which entitles the contractor to take steps to terminate the contract under the appropriate clauses (SBC and IC: clause 8.9, MW clause 6.8). Failure to pay three certificates probably entitles the contractor to accept the employer's conduct as repudiatory, bring the obligations of both parties to a permanent end and claim damages.[28] But if there has been no acceptance of the repudiation and the contractor has merely suspended its obligations, it amounts to a breach of contract on the part of the contractor. It is probably not a breach that entitles the employer to accept it as repudiation for two reasons:

- The contractor is not saying that it will never continue its obligations, but simply that it will suspend them until it is paid. An expressed intention to suspend precludes the implication of an intention to bring the contract to an end.[29]
- The contractor intends to comply with the contract, albeit it has gone about this improperly, therefore there is no intention to repudiate the contract.[30]

Nevertheless, it is a breach of contract and the employer would be entitled to such damages as flow from the breach. The cost of rectifying the acts of vandalism might well fall into that category.

The position is totally different if the contractor has given the requisite notice. When the notice expires, the contractor is entitled to leave the site. The contractor owes a general duty to leave the site in a safe condition, but the duty to insure comes to an end together with all the contractor's other contractual duties. If vandalism occurs, the failure of the employer to make payment is not the cause, but it is a circumstance without which the vandalism probably would not have occurred. The reason the site was unprotected

28 D R Bradley (Cable Jointing) Ltd v Jefco Mechanical Services (1989) 6-CLD-07–19

29 F Treliving & Co Ltd v Simplex Time Recorder Co UK Ltd (1981) unreported

30 Woodar Investment Development v Wimpey Construction UK [1980] 1 All ER 571

was because the contractor had exercised the right to suspend. The problem would belong to the employer, along with all the other problems associated with total cessation of work on site.

21   In DB, if the employer provides a site investigation report and the ground conditions are found to be different, who pays any extra cost?

Under the provisions of DB, the Employer's Requirements set out the criteria that the contractor must satisfy in preparing the Contractor's Proposals. That the Contractor's Proposals are a response to the Employer's Requirement is made clear by the second recital. The contractor is likely to be able to found a claim against the employer if site conditions are not as assumed in the Employer's Requirements.[31]

If the employer makes a statement of fact, intending that the contractor will act upon it, it is a 'representation'. Such statements are often made in tender documents. Many of the statements made by the employer within the Employer's Requirements will be representations. The contractor will use the information in compiling its tender. Typically, this will include information about the site and ground conditions. If any of the statements of fact are incorrect, they will probably amount to misrepresentations. A misrepresentation is of no importance unless it is made part of the contract or if it was an inducement to the contractor to enter into the contract. In such cases, the contractor may well have a claim for damages at the least. The precise remedies available depend upon whether the misrepresentation is fraudulent, negligent, innocent or under statute.

The remedies used to be restricted to cases of fraud or recklessness but, as a result of the Misrepresentation Act 1967, they now apply to all misrepresentations. The onus lies with the party who made the representation to prove reasonable ground for believing and actual belief, up to the time the contract was made, that the facts represented were true. Section 3 of that Act restricts the employer's power to exclude liability for misrepresentation.

It follows that a contractor may have a claim for misrepresentations about site conditions made during pre-contractual

---

31   *C Bryant & Son Ltd v Birmingham Hospital Saturday Fund* [1938] 1All ER 503

negotiations. It may claim for damages for negligent misrepresentations or breach of warranty or under the Misrepresentation Act 1967 arising out of representations made or warranties given by or on behalf of the employer.[32] However, the efficacy of all claims depends upon circumstances. In an Australian case (*Morrison-Knudsen International Co Inc v Commonwealth of Australia* [33]), the contractor claimed that information provided at pre-tender stage 'as to the soil and its contents at the site ... was false, inaccurate and misleading ... the clays at the site, contrary to the information, contained large quantities of cobbles'. There was a trial of a preliminary issue and it was concluded that the basic information in the site document appeared to have been the result of a great deal of technical effort on the part of a department of the defendant. It was information that the plaintiffs had neither the time nor the opportunity to obtain for themselves. It might be doubted whether they could have been expected to obtain it by their own efforts in their role as a tenderer. But it was crucial information for the purpose of deciding the work to be carried out.

A representation followed by a warning that the information given may not be accurate will not usually be sufficient to protect the employer, because it is a clear intention to circumvent section 3 of the Act. Indeed, such a statement may convert the representation into a misrepresentation.[34]

A misrepresentation may also amount to a collateral warranty. For example, in *Bacal Construction (Midland) Ltd v Northampton Development Corporation,*[35] which involved a design and build contract, the contractor was instructed to design foundations on the basis that the soil conditions would be as set out in borehole data provided by the employer. The Court of Appeal held that there was a collateral warranty that the ground conditions would be in accordance with the basis on which Bacal had been instructed to design the foundations and that they were held entitled to damages for its breach.

However, care must be taken, because it has been held that a contractor in a design and build situation is not entitled to rely on

32 *Holland Hannen & Cubitts (Northern) Ltd v Welsh Health Technical Organisation* (1981) 18 BLR 80
33 (1972) 13 BLR 114
34 *Cremdean Properties Ltd v Nash* (1977) 244 EG 547
35 (1976) 8 BLR 88

a ground investigation report that is simply made known to the contractor by a reference to it on a drawing. It was held not to be incorporated into the contract but had been noted simply to identify a source of relevant information for the contractor. In the court's view, that was not sufficient to override a clause in the contract that placed on the contractor the obligation to satisfy itself about the nature of the site and the subsoil.[36] In addition, the judge made the following thought-provoking observation:

> The nature of a ground investigation report is such that it is unlikely, it seems to me, that parties to a contract would wish to incorporate it into a contract between them. All it can show is what was the result of particular soil investigations. If parties did in fact seek to incorporate a ground investigation report into a contract between them, difficulties could arise as to what was the effect in law of so doing. If one has regard to the terms of the [site investigation] report, and in particular to those two parts of the narrative in which there is a reference to ground water, it really is impossible to say that any definite or positive statement of a nature such as could amount to any sort of contractual term was made. Rather, the information given was hedged about with qualifications as to the accuracy and reliability of what was shown by the investigations undertaken.

Quite so. Nevertheless, employers continue to include site investigation reports as part of the contract information given to the contractor before tendering takes place.

36 *Co-operative Insurance Society v Henry Boot Scotland Ltd* (2003) 19 Const LJ 109

3 Contractor's programme

22 If an architect approves a contractor's programme, can the contractor subsequently change the programme without the architect's knowledge and, if so, can the architect demand an update?

Provision of a master programme by the contractor is covered by clause 2.9.1.2 of SBC, which requires the contractor to provide two copies to the architect. Approval of the programme by the architect is dealt with in the next question. Suffice to say that it has no particular effect on the contract or contractor's responsibility for the programme.

SBC is the only one of the JCT traditional contracts that actually refers to the contractor's programme. Clause 2.9.1.2 should be carefully studied. It states that the contractor will provide the architect without charge with two copies of the master programme for the Works. The contractor's obligation to revise the programme only occurs if an extension of time is given either by the architect under clause 2.28 in the normal way or by a pre-agreed adjustment as a result of the acceptance of a schedule 2 quotation.

The contractor has no obligation to revise the programme if the contractor falls behind due to its own fault. Looked at stringently, that makes perfect sense. If it is the contractor's fault that a delay has occurred, it is clearly the contractor's responsibility to recover the lost time under clause 2.28.6.1, which requires the use of best endeavours *constantly* to prevent delay. It can be convincingly argued that the programme requires no adjustment because the contractor ought to be doing everything a prudent contractor would do to get back on programme. The only time the programme requires adjustment is if the completion date is adjusted to a later date. In that case, the programme should be amended to reflect completion on that later date.

The practice of contractors constantly submitting revised programmes, not because the completion date has been revised but because the contractor has fallen behind, is to be deplored. The architect should not be interested in when the contractor says it believes it will finish if that date is not the completion date in the contract. The problem is that many contractors take the view that the contractual completion date is simply a date to aim for and that, if it is not achieved, the architect will probably, if sufficiently threatened, extend the completion date to match the date of practical completion. The constant submission of programmes that bear no relation to the contract or extended date for completion does nothing to assist the architect in considering extensions of time or disruptions to the regular progress of the work. For that purpose, a programme that accurately reflects the intended progress and completion is essential.

All the foregoing is by way of putting the question into context. Clause 2.9.1.2 is the only clause that refers to the contractor's programme and it will readily be seen that nothing in the clause states that the contractor must comply with its own programme. First, it should be noted that the reference is to a master programme. That allows the contractor to produce numerous detailed programmes which there is no obligation to provide for the architect. Second, the contractor can opt not to work to its own programme or even to change the programme without informing the architect. Only if the change results from an extension of time must the revised programme be submitted. Therefore, it is possible for a contractor to submit a programme indicating that work will progress from point A through B, C, D etc. to Z and, on site, commence at point Z and work in reverse. There is nothing in the contract that obliges the contractor to work to its programme. Most contractors do, of course, but the question was no doubt prompted by the fact that from time to time a contractor will find it useful to significantly vary its work from the submitted programme.

The programme is not a contract document, although it might be termed a contractual document, because it is generated in accordance with a clause in the contract. However, neither the contractor nor the employer is bound to follow it. It would be possible to make the programme a contract document, but that would not necessarily be an advantage, because every slight deviation from the programme potentially would have a financial implication.[37]

Therefore, as the standard contract currently stands, the contractor can change the programme without the architect's knowledge or permission and the architect has no power to require an update. Obviously, additional clauses can be introduced to deal with some of these difficulties and programmes can be required in particular formats. However, considerable thought should be given before deciding to make it a contract obligation for the contractor to comply with its own programme. There could be substantial financial repercussions as noted above.

23 Under SBC, the architect has approved the contractor's programme, which shows completion 2 months before the contract completion date. Must the architect work towards this new date?

It is quite common for a contractor to show a date for completion on the programme that is before the date for completion in the contract. It is never very clear why a contractor should do this, because the programme cannot alter the contract date for completion. All the programme does is to inform the employer and architect that the contractor intends to complete before the completion date. The contractor is entitled to do this because clause 2.4 states that it must complete the Works 'on or before' the completion date.

By asking whether the architect must 'work towards this new date', the questioner can be referring only to three important situations: the most obvious is the provision of further information to the contractor; another is responding to a notice of delay under clause 2.28 where the architect, in deciding whether to give an extension of time, is to consider whether completion of the Works is likely to be delayed beyond the completion date; the final important one is the architect's obligation under the same clause 2.28 to notify the contractor about an extension of time decision no later than the completion date. Other matters relate to the date of practical completion, which ought to be the same as, but which is not

---

37 *Yorkshire Water Authority v Sir Alfred McAlpine and Son (Northern) Ltd* (1985) 32 BLR 114

necessarily connected to the contract completion date. The contract date for completion is the date by which the contractor undertakes to complete the whole of the Works. The date of practical completion is the date by which virtually the whole of the Works are, as a matter of fact, complete.

To deal with the provision of information first, the courts have held that, although the contractor is entitled to complete before the completion date in the contract, the employer has no obligation to assist the contractor to do so.[38] This principle is now enshrined in the contract at clause 2.12.2. This clause stipulates that, in providing further information, the architect must have regard to the progress of the Works. This means that if the contractor is progressing so as to complete by the completion date, the information must be provided at such times as will enable the contractor to finish the Works on time; but that if the contractor is making slow progress, the architect is entitled to slow the delivery of information to suit. However, the clause proceeds to state that, if the contractor seems likely to finish before the completion date, the architect need only provide information to enable it to complete by the completion date.

The second and third situations deal with extension of time and assume that everyone knows the completion date. The only way in which the architect's responsibilities under clause 2.28 can be affected is if the contractor's programme, showing a completion 2 months earlier than the contract date, somehow takes precedence over the completion date in the contract. If that is the case, it will clearly affect the provision of information also.

The contract date for completion, like the other terms of the contract, can be changed only by agreement between the parties – the employer and the contractor. Approval of the programme by the architect is of little or no consequence. The programme is not a contract document and the parties are not bound by it – not even the contractor, strange as that may seem. Most architects will not approve the contractor's programme, but even if the programme is approved, it merely signifies that the architect is happy with it. In effect the architect is saying: 'I am happy for you to work in accordance with the programme if that is what you wish to do.' Approval does not transfer the responsibility for the thing

38 *Glenlion Construction Ltd v The Guinness Trust* (1987) 39 BLR 89

approved from the contractor to the architect[39] unless the contract makes specific provision for that to happen.

The position might be different if the contractor submitted its programme showing completion two months earlier than the contractual date and it was discussed and agreed by both parties and the new date confirmed. In that instance, the architect may be obliged to work in every way as though the programme date for completion was the contract date for completion. The confirmed agreement of both parties amounts to a variation of the contract terms. In practice, this could take place at a site meeting; possibly the one usually and incorrectly called the 'pre-contract meeting' ('pre-start meeting' is a better name). Where the parties reach an agreement at such a meeting, with plenty of witnesses and where the agreement is recorded in the minutes which are subsequently agreed by all parties, it is difficult to escape the conclusion that a valid variation of the contract terms has taken place.

If the circumstances are such that a valid variation of the terms has not occurred and it is only the architect who has received and commented on the programme, the contractor's obligation is still to complete by the contract date for completion. As noted earlier, it is difficult to see what the contractor gains by this strategy. If the contractor notifies delay, the architect appears to be able to take the contractor's proposed completion date into account when deciding whether completion of the Works is likely to be delayed beyond the completion date. That would not work to the contractor's advantage. Suppose the contract completion date was 30 September and the contractor's proposed date for completion was 30 July. If the contractor argues that it is being delayed in completion by 2 weeks, there seems to be no reason why the architect should not assume that the contractor will, therefore, complete by the middle of August. That would not indicate a delay to the contract completion date and no extension of time would be due. On this basis, the contractor would have to register a delay of more than 2 months before an extension of time would be due.

---

39 *Hampshire County Council v Stanley Hugh Leach Ltd* (1991) 8-CLD-07–12

24   Can the architect insist that the contractor submit the programme in electronic format?

The rights and obligations of the employer and the contractor are governed by the terms of the contract. None of the standards forms in general use require the contractor to submit a programme in electronic format. Indeed, only one of the JCT series of contracts requires the contractor to submit a programme at all and that is SBC, which requires the contractor to submit a master programme to the architect.

There is no doubt that a programme in electronic format can greatly assist the architect in deciding upon the correct extensions of time, in considering applications for loss and/or expense and in generally monitoring the progress of the Works. Of course, there are a number of different computer programs that will support a building programme and it is essential that any programme submitted by a contractor in electronic format is compatible with the software on the architect's computer. The alternative is to ask the contractor to supply a list of the activities on the programme together with the relevant predecessors and successors so that they can be entered into the architect's computer. This is a rather more tedious operation than simply inserting a disk, but it is far better than having no such programme at all.

All that is necessary to require the contractor to submit an electronic formatted programme is for a suitable clause to be inserted in the preliminaries section of the bill of quantities or specification. This will not conflict with the printed contract in most cases because, as noted earlier, only SBC requires a programme of any kind. It will not conflict with SBC, because the contract does not specify the type of programme, therefore the additional clause is not overriding or modifying the printed form but merely amplifying it by adding further requirements. In this clause, the architect can require the programme in a form to suit his or her own computer software. A contractor who fails to supply the programme in the form specified will be in breach of contract and the architect is entitled to take account of the breach in calculating extensions of time.[40]

The use of computers in dealing with extensions of time has been approved by the courts. What is the position if there is no requirement for the contractor to supply an electronic version of

---

40 *London Borough of Merton v Stanley Hugh Leach Ltd* (1985) 32 BLR 51

the programme? Can the architect compel the contractor to provide it in any event? The answer to that appears to be that, in the absence of a specific clause in the contract, the architect cannot compel the contractor to provide a programme in any particular form. However, when a contractor notifies delay in the expectation of receiving an extension of time, there is nothing to prevent the architect asking for a programme in a specific form as part of the architect's general power to request further information from the contractor. The contractor may refuse to provide it, arguing that there is nothing in the contract that requires it. The architect's reply would be to simply point out to the contractor that, in the absence of such a programme as requested, the architect will find it very difficult to determine a fair and reasonable extension of time and that the contractor will be the author of its own misfortune if it does not receive the extension of time it expects. That usually has the desired effect. If not, the architect may give a somewhat shorter extension of time than expected and, if this is done as part of the review of extension under SBC, IC or ICD during the 12-week period after practical completion, the contractor will have to live with it, because the architect has no power under these contracts to revisit the situation again once the 12 weeks has expired. Indeed, under MW and MWD, there is no review provision and, therefore, no opportunity for the architect to reconsider after the date for completion in the contract, or as already extended, has passed.

25 When a contractor says that it owns the float, what does that mean?

Float is a commonly used term that is much misunderstood. It is the difference between the period required to perform a task and the period available in which to do it. Critical activities have no float – that is why they are critical: there is no difference between the period needed to carry out the activity and the period allocated for it. In other words, there is no scope for any delay at all before the completion date of the project is affected. To say that an activity has a day of float means that the activity could be extended by another day without affecting the completion date of the project. Put another way, the activity could commence a day late without there being any effect on the overall programme.

Contractors sometimes argue that an extension of time is due even if a non-critical activity is delayed. They argue that they own the float and, therefore, that the employer cannot 'take advantage' of it. This is a very strange contention. No one owns float. It is like trying to argue that a person is taking advantage of the air around them. Float is simply the space before or after individual activities when they are put together in the form of a programme. Whether it actually exists at all depends on whether the programme is accurate.

It is sometimes argued that if a contractor programmes to complete a 12-month contract in 10 months, the 2 months are the contractor's float and, therefore, if the project is delayed by even a week, an extension of time will be due, even if the contractor finishes several weeks before the completion date. That is obviously wrong. If the week's delay causes float to be used, because of some employer default, the contractor has no entitlement to an extension of time because the delay does not affect the completion date; that is the crucial point.

In *Ascon Construction Ltd v Alfred McAlpine Construction Isle of Man Ltd*,[41] the judge put the position like this when dealing with a contractor's claim against a sub-contractor:

> ... I must deal with the point made by McAlpine as to the effect of its main contract 'float' ... It does not seem to be in dispute that McAlpine's programme contained a 'float' of five weeks in the sense, as I understand it, that had work started on time and had all sub-programmes for sub-contract works and for elements to be carried out by McAlpine's own labour been fulfilled without slippage the main contract would have been completed five weeks early. McAlpine's argument seems to be that it is entitled to the 'benefit' or 'value' of this float and can therefore use it at its option to 'cancel' or reduce delays for which it or other sub-contractors would be responsible in preference to those chargeable to Ascon.
>
> In my judgment that argument is misconceived. The float is certainly of value to the main contractor in the sense that delays of up to that total amount, however caused, can be accommodated without involving him in liability for liquidated damages to the employer or, if he calculates his own prolongation costs from the contractual completion date (as McAlpine

has here) rather than from the earlier date which might have been achieved, in any such costs. He cannot, however, while accepting that benefit as against the employer, claim against the sub-contractor as if it did not exist. That is self-evident if total delays as against sub-programmes do not exceed the float. The main contractor, not having suffered any loss of the above kinds, cannot recover from sub-contractors the hypothetical loss he would have suffered had the float not existed, and that will be so whether the delay is wholly the fault of one sub-contractor, or wholly that of the main contractor himself, or spread in varying degrees between several sub-contractors and the main contractor. No doubt those different situations can be described, in a sense, as ones in which the 'benefit' of the float has accrued to the defaulting party or parties, but no one could suppose that the main contractor has, or should have, any power to alter the result so as to shift that 'benefit' The issues in any claim against a sub-contractor remain simply breach, loss and causation.

I do not see why that analysis should not still hold good if the constituent delays more than use up the float, so that completion is late. Six sub-contractors, each responsible for a week's delay, will have caused no loss if there is a six weeks' float. They are equally at fault, and equally share in the 'benefit'. If the float is only five weeks, so that completion is a week late, the same principle should operate; they are equally at fault, should equally share in the reduced 'benefit' and therefore equally in responsibility for the one week's loss. The allocation should not be in the gift of the main contractor.

I therefore reject McAlpine's 'float' argument.

This is good authority that float is owned by no one. The decision in *How Engineering Services Ltd v Lindner Ceilings Partitions plc* is to similar effect.[42] Therefore, when a contractor says that it owns the float, it means that the contractor does not properly understand the concept of float.

42 17 May 1995 unreported

4 Contract administration

26 Does the contractor have a duty to draw attention to an error on the architect's drawing?

Generally and in normal circumstances, the contractor has no liability for design, therefore no liability for the production of design drawings. The question often arises whether the contractor is entitled simply to build what the drawings and specifications set out, even if there are errors on the architect's drawing. It is not surprising that most architects would say no, but the case law on this subject is not so clear.

It has been established by a Canadian case that a contractor will be liable to the employer for building errors in a design if the original architect was not involved in the construction stage.[43] The *ratio* of that case seems to have been that the employer was no longer relying on the architect and, therefore, relied solely on the contractor, which should have taken care to check that everything on the original architect's drawing worked properly. That, however, is not the situation under consideration here, where the original architect is still engaged, but where there is an error in the drawings. There were two cases in 1984[44] which held that a contractor did have a duty to warn the architect if it believed that there was a serious defect in the design. Subsequently, however, another court decided that such duty as the contractor might have was to the employer and probably only in those cases

---

43 *Brunswick Construction v Nolan* (1975) 21 BLR 27
44 *Equitable Debenture Assets Corporation Ltd v William Moss* (1984) 2 Con LR 1; *Victoria University of Manchester v Wilson and Womersley and Pochin (Contractors) Ltd* (1984) 2 Con LR 43

where the contractor was aware of the employer's reliance for at least part of the design.[45] This has echoes of the Canadian case mentioned on page 46. To further confuse matters, another case held that a contractor had a duty to at least raise doubts with the architect if there appeared to be something wrong with the drawings.[46] One would have to wonder at the motives of a contractor who had full knowledge of a drawing error and yet failed to draw it to the attention of the architect.

That position was taken a stage further by a Court of Appeal case.[47] Although this case involved sub-contract work, the principles set out by the court are equally applicable to main contracts. JMH designed the temporary support work to a roof. Unfortunately, its design was overruled by the employer's engineer, who proposed a different design. There was no question in this instance over whether JMH failed to warn the engineer. They did warn the engineer of the danger of his design quite clearly, but he took no notice and the engineer's design for temporary work went ahead. Needless to say, the roof collapsed. Surprisingly, the court held, not just that JMH had a duty to warn, which the court seemed to accept had been done, but that they had failed to warn with sufficient force. One cannot help but think that the only degree of warning that the court would have accepted as sufficient would have been if JMH had given the warning and, at the same time, threatened to stop work if the warning went unheeded. This appeared to be the court's position also.

The contractor's duty to warn probably arises only if the design is seriously defective. In the case just mentioned, it seems to have been a potential danger to life. A contractor who did not warn an architect who had made a small dimensional error or a small mistake in detailing would be unlikely to have any liability.

The important point to be drawn from these cases is the reliance by the employer on the contractor. If it can be shown that the employer does rely, even partly, on the contractor, it seems that there will be a duty to warn of serious defects. On the other hand,

45 *University Court of the University of Glasgow v William Whitfield & John Laing (Construction) Ltd* (1988) 42 BLR 66

46 *Edward Lindenberg v Joe Canning & Jerome Contracting Ltd* (1992) 9-CLD-05–21

47 *Plant Construction plc v Clive Adams Associates and JMH Construction Services Ltd* [2000] BLR 137

cases where the duty arises to warn the architect will be rare, because the architect seldom, if ever, relies or is entitled to rely on the contractor. In the context of JCT traditional contracts, the duty is likely to be limited, because the employer will usually be relying on the architect and not the contractor. Contractors can take heart that they are not generally responsible for checking the architect's drawings. Having said that, a contractor that proceeded with construction in the certain knowledge that there were errors on the drawings would find little favour with an adjudicator in any subsequent dispute.

27   Under DB, must the employer's agent approve the contractor's drawings?

Clause 2.8 provides that the contractor must without charge give the employer two copies of its design documents as and when from time to time necessary and in accordance with schedule 1 of the contract or as otherwise stated in the contract documents. The contractor is not to commence any work until it has complied with the procedure.

Schedule 1 sets out the procedure, but with reference to the Employer's Requirements. Paragraph 1 requires submission in the format stated in the Employer's Requirements. Therefore, if the Employer's Requirements do not state the format, it seems the contractor may submit the information in any format it desires. It might even be argued that, in the absence of a stated format, the contractor effectively need not submit at all. That is a very strict view, but one which a contractor is entitled to take. Therefore, it is essential that the format is set out.

The submission must be made in sufficient time to allow any comments made by the employer to be incorporated before use of the relevant document. That must be read in the context of paragraph 2, which gives the employer 14 days from receipt of the submission or, if the contract documents give a later date or a period, from the date or the expiry of the period, to return one copy of the document to the contractor.

The contract adopts the well-known system of lettering the returned documents either 'A', 'B' or 'C', depending on whether or not they are in accordance with the contract. 'A' means that the contractor must carry out the Works in accordance with that document. 'B' or 'C' means that the document is not in accordance with the contract and it must be accompanied by a written statement stating why the employer considers that to be the case. Documents marked 'B' may be used by the contractor if the employer's comments are incorporated and the employer is provided with an amended copy. Documents marked 'C' cannot be used for construction, but the contractor may resubmit after amendment.

If the contractor thinks that the employer is wrong and that the document is in accordance with the contract, there is the option under paragraph 7 of notifying the employer within 7 days of receipt of the comment that compliance with the comment will result in a change (i.e. a variation). The contractor must give a reason, of course. The employer has a further 7 days to either confirm or withdraw the comment. If the employer simply confirms the comment, the contractor must then amend and resubmit the document. Paragraph 8 then sets out some provisos:

- Whether the employer confirms or withdraws comments does not mean that the employer accepts that the documents or amended documents are in accordance with the contract or that compliance with the comments will result in a change.
- If the contractor does not take the option of notifying the employer that compliance with the comment will result in a change, the comment is not to be treated as giving rise to a change.
- The contractor's duty to ensure that the design documents are in accordance with the contract is not reduced by the contractor's compliance with the submission procedure or with the employer's comments.

In brief the position is that it is the contractor's obligation to comply with the contract. No submission of documents or comments by the employer will remove that obligation. If the employer makes comments that amount to a change, the contractor must promptly notify the employer of its view on the matter. Failure to notify the employer within the 7 days allotted will preclude the contractor from recovering any payment for such alleged change. However, notification, in itself, will not guarantee payment; it will be a matter of fact whether or not there has been a change.

It should be noted that the contract stays well clear of any suggestion that the employer approves any documents. But use of the word 'approval' appears not to make any difference to the principle in any event. In *Hampshire County Council v Stanley Hugh Leach Ltd*,[48] the court said:

> The fact that Leach's alternative proposals were approved by the architects is irrelevant. No employer is going to be advised to enter into a contract giving the contractor an entirely free hand.
>
> The JCT Design and Build Contracts require the contractor's design be approved and this of course does not relieve the contractor of obligations in respect of his design.

28   What happens if the contractor cannot obtain materials?

The authority on this topic is scarce to the point of non-existence.

Under SBC clause 2.3.1, materials and goods have to be provided only 'so far as procurable'. It will be noticed, however, that the contractor's obligation under clause 2.1 is to provide what is specified in the contract documents. The whole of the contract must be read together, of course, and the introduction of the word 'procurable' gives the contractor a useful protection if materials or goods are truly unobtainable. Clearly, it does not protect a contractor who discovers that it has miscalculated its tender and that it is more difficult or more expensive than expected to provide what is specified. 'Procurable' is not qualified and, on a strict reading of the clause, it can even be argued that a contractor is protected even if the materials or goods were not procurable before the contract was entered into. It might be thought that a sensible and businesslike approach would restrict the meaning of 'procurable' to those items that had become unobtainable after the contract was executed. Whether that is the correct way to interpret it is not certain.

It is difficult to forecast what conclusion might be reached by the courts, still less an adjudicator, but a strict reading of the clause results in the conclusion (deeply unattractive so far as the employer is concerned) that if the items are not procurable for any reason, the contractor's obligation to provide them is at an end. It then becomes necessary for the architect to issue an architect's instruction requiring as a variation the provision of a substitute material. The variation is to be valued in the usual way.

The effect is to remove from the contractor any obligation to check that specified goods and materials are procurable before tendering. In order to change this situation, it is probably necessary to amend the contract clauses to specify a date after which the contractor is not responsible if materials or goods are no longer procurable.

It is arguable that if the contract does not refer to materials being procurable, the contractor's inability, through no fault of its own, to obtain specified materials may render the contract frustrated.

29 What powers does a project manager have in relation to a project?

Over the last few years the concept of project management has steadily gained ground, together with a good many misconceptions. A project manager is unlikely to be the same person as the contract administrator and it is the contract administrator who has the main powers under the building contract. The RIBA approved a definition of a project manager as follows:

> The Project Manager is a construction professional who can be given *executive authority and responsibility* to assist the client to identify the project objectives and subsequently supply the technical expertise to assess, procure, monitor and control the external resources required to achieve those objectives, defined in terms of time, cost, quality and function.

It may be argued that such a definition does nothing to set out what ought to be the function of a project manager in regard to a building project. Project management is often considered as though it is a self-contained system and that the words 'project manager' instantly conjure up a recognisable and easily identifiable discipline. Even the courts have agreed that this is not the case and that the duties of a project manager may vary dependent on the base discipline of the person carrying out the role.[49] The concept of project management is not particularly linked to construction; a project manager is rather a creature of the manufacturing industries. It is certain that all project managers have skills in common, but a project manager on a building contract

49 *Pride Valley Foods Ltd v Hall & Partners* (2000) 16 Const LJ 424

cannot approach the task with the same freedom as if he or she were project managing a new product through a factory. There are roughly two kinds of project managers:

• project managers who represent the employer and act as its technical arm
• project managers who not only represent the employer but also carry out the contract administration role in regard to building contracts.

The first type of project manager acts as the employer's representative and generally acts as agent for the employer with the power to do everything the employer could do in relation to the project. This is probably the usual position occupied by the person termed project manager. He or she will appoint consultants and carry out the briefing exercise having first been briefed by the employer, make the final decision about the selection of a building contractor and answer any queries from the professional team. The theory is that the employer has the benefit of a skilled professional looking after his or her interests and being paid to watch the other professionals. This kind of project manager has no powers under the building contract although he or she may try to enter site, chair site meetings and give instructions directly to the contractor. Some project managers even insist on countersigning all certificates. That kind of activity on the part of the project manager, though regrettably common, is unlawful and it may lead to disputes. It always leads to confusion. A contractor taking instructions from such a project manager is most unwise.

Most building contracts do not even acknowledge the project manager's existence. There is provision under SBC for an employer's representative, who might well be the project manager, to be appointed to carry out the employer's functions. Some other contracts, for example GC/Works/1 (1998) refer to the project manager, but they might as well have used the phrase 'contract administrator' and simply add to the current confusion over the role.

The second type of project manager performs all the functions of a contract administrator so far as the building contract is concerned. The project manager in this situation has a great deal of power, because acting as the employer's representative is added to the role. The project manager's function is commonly thought by employers to be the management of the project and this type of project manager is closest to that situation. It is fairly unusual to find a project manager in this role. That is possibly just as well, because it is tantamount to having the employer as contract administrator and there is no properly independent professional for the issue of certificates. That very much devalues any certificates and makes them no more than the employer's view, which carries no more weight than the contractor's view.[50]

It is possible to find a project manager working solely for a contractor. In such cases, the project manager has no more, although different, power than a project manager acting solely for the employer. Indeed, it is useful to compare them.

---

50 *J J Finnegan Ltd v Ford Sellar Morris* (No.1) (1991) 25 Con LR 89

5 Architects

30 Planning permission was obtained for a small building. The building owner wants to press ahead with a larger building without further reference to Planning. The architect knows that the Planning Department would refuse the large building out of hand. Should the architect continue to do the drawings and administer the contract on site?

It is the architect's duty to advise the client on all aspects of the building process about which the architect professes expertise. Obviously, town planning is one area where the architect should have expertise – not the expertise of a town planner or of an expert planning consultant or of a lawyer skilled in this area of the law, but certainly the ordinary aspects of town planning that one would normally expect the architect to know as part of the general architectural skills.[51]

In this case, planning permission had been obtained for a particular building. The client had clearly had a change of mind and wanted a bigger building on the same site. There is nothing wrong with that, but architects have a duty to advise the client about seeking planning permission again in such circumstances. If intending to press ahead without obtaining planning permission, the client is going to do something unlawful and the architect has a duty to make that crystal clear. In addition, if the architect knows for certain that making a planning submission would be pointless, there is an added duty to convey this to the client in the strongest terms.

51 *B L Holdings v Robert J Wood and Partners* (1979) 12 BLR 1

The architect is being asked to finish the drawings for the larger building that would not gain planning approval even if application were made. The architect knows that the purpose for which the drawings are intended is the unlawful construction of a building. Architects placed in this position should flatly refuse to have anything further to do with the project. This would not amount to repudiation at law, much less be in breach of any part of the professional Code of Conduct. Quite the contrary: the larger building is not something for which the architect was engaged. In other words, it is not part of the terms of engagement (for the smaller building for which planning permission has been obtained) and, therefore, the architect cannot be in breach by refusing to do the work.

For an architect to collaborate with the client in enabling the construction of a building for which it is known that planning permission would be refused is an unlawful act and one that is contrary to the Code of Conduct. If the architect did not know that planning permission was not to be sought or that it could not be obtained, there would be no wrongful behaviour until the architect knew, or should have known, the true situation. On becoming aware of the true situation, if advice to the client fell on deaf ears, the architect would have no option but to stop work.

31 SBC. Roof tiles were varied at site meeting and by letter, but the contractor ordered the original ones. Should the architect have changed the drawings as well?

Under SBC, clause 3.12.1, all instructions are to be in writing. That means that instructions can be issued in the form of a letter or on a specially designed RIBA form or on the architect's specially designed form or on anything at all, provided that it is in writing. An instruction must also be expressed positively. It must instruct the contractor about something and it must be empowered by the contract.

SBC states what is included in the contract sum in clause 4.1. If there are bills of quantities, the quality and quantity of the work is deemed to be what is set out in the bills. If there are no bills but there are some quantities, the quality and quantity is deemed to be what is contained in the items with quantities; otherwise the drawings take precedence. But if there is no conflict, the drawings and specifications and schedules of work are to be read together.

However, this is simply what is in the contract sum. When an instruction is issued requiring a variation under clause 3.14, there is no requirement for a drawing, bills of quantities or a specification. All that is required is that the content of the instruction is clear to the contractor. Usually the architect will either give a very full specification or a drawing.

In this instance, the roof tiles were varied at a site meeting and by letter. The variation at the site meeting falls into the category of oral instructions, which require either architect or contractor confirmation within 7 days. The architect can always ratify it later if nobody confirms. The point is that, failing a written confirmation, the instruction does not become effective until the architect ratifies it either by issuing minutes of the meeting that have been drafted by the architect or by a letter. Nothing in the contract requires that the existing drawings must be amended to suit. Indeed, it is sometimes best if amendments are not made to the drawings, because then the effect of the variation can more easily be seen subsequently. The ordering of the original tiles amounts to a failure on the part of the contractor if the written instruction or confirmation of the site meeting instruction was received by the contractor before the order was sent.

32   Are there any circumstances in which a contractor can successfully claim against the architect?

This is a question that crops up fairly frequently. Architects are prone to ask it just before making an important contractual decision; contractors ask it when they are particularly annoyed with an architect's conduct. In general terms, it is usually easier for the contractor to claim against the employer than the architect. That is because the contractor and the employer are related by the building contract while the contractor has no contractual relationship with the architect. Therefore, a contractor finds it relatively easy to claim under the terms of the contract or even against the employer for a breach of the contract. Claims by the contractor against the architect have to be in tort. Since the case of *Murphy v Brentwood District Council*,[52] negligence claims have become very difficult to sustain. A contractor making a claim against an architect would almost certainly do so under the reliance principle.[53] It usually applies to professionals, although courts have extended the scope in some instances. The principle, in brief, is to the effect that if a professional gives advice to another person or class of people knowing that the person or persons will rely on it, and if the person or persons does rely on the advice and, as a result, suffers a loss, the loss will be recoverable from the professional. This is irrespective of any fee paid or not paid to the professional and even if there is no contractual relationship.

The contractor sued both employer and architect in *Michael Salliss & Co Ltd v ECA Calil,*[54] claiming that the architects owed a duty of care to the contractor. Although the contractor was unsuccessful in arguing that the architect owed the contractor a duty to provide accurate and workable drawings, it was successful in its claim that it relied on the architect to grant an adequate extension of time and properly to certify the value of work done. The court appeared to think that it was self-evident. It remarked that, if the architect unfairly promoted the employer's interest by inadequate certification or merely failed properly to exercise reasonable care and skill in the certification, it was reasonable that the contractor should not only have rights against the owner but also against the architect to recover damages.

Three years later, *Pacific Associates v Baxter*[55] seemed to overturn this position, saying that a question mark hung over the *Sallis* case. Pacific Associates was effectively the contractor under a FIDIC contract for work in Dubai. The contractor claimed that it had encountered unexpectedly hard materials and that it was therefore entitled to a substantial extra payment. The engineers would not certify the amount claimed and the contractor sued them. The claim alleged that the engineers acted negligently in breach of their duty to act fairly and impartially in administering the contract. In a judgment upheld by the Court of Appeal the court struck out the claim, on the basis that the contractor had no cause of action or, in other words, that it could not make a claim in that particular way. In making that decision, the court was mindful that there was provision for arbitration between employer and contractor and, therefore, the contractor could have sought arbitration on the dispute.

---

53 *Hedley Byrne & Co Ltd v Heller & Partners Ltd* [1963] 2 All ER 575
54 (1986) 6 Con LR 85
55 (1988) 44 BLR 33

The court also referred to a special exclusion of liability clause in the contract by which the employer was not to hold the engineers personally liable for acts or obligations under the contract, or answerable for any default or omission on the part of the employer. The fact that the engineers were not a party to the FIDIC contract – just as architects are not parties to JCT contracts – seems to have been ignored by the court.

The question mark appears to properly hang over the *Pacific Associates* case rather than the *Sallis* case. Whether a duty of care exists does not depend on the existence of an exclusion of liability clause, except to the extent that the existence of such a clause suggests acceptance by the engineer that there is a duty of care which, without such a clause, would give rise to the liability. The clause in question might well be deemed unreasonable under the provisions of the Unfair Contract Terms Act 1977.[56] Why the inclusion of an arbitration clause should exclude engineers from liability to the contractor is not immediately (or even subsequently) obvious. The fact that the parties chose to settle their disputes by arbitration cannot excuse the engineers from their duty to both parties.

In *Lubenham Fidelities v South Pembrokeshire District Council*,[57] the Court of Appeal (by which the court in the *Pacific Associates* case should have been bound) expressly confirmed that the architect owed a duty to the contractor in certifying. The architects in that case were not held liable, but that was because the chain of causation was broken and the contractor's damage was caused by its own breach in wrongfully withdrawing from site. But the court reached its conclusion with reluctance, because the architects' negligence was the source from which the sequence of events began to flow. The court expressly stated that, because the architects were appointed under the contract, they owed a duty to the contractor as well as to the employer to exercise reasonable care in issuing certificates and in administering the contracts correctly and that, by issuing defective certificates and in advising the employer as they did, the architects acted in breach of their duty to the contractor.

The court was simply following the precedent of earlier courts. In *Campbell v Edwards*,[58] the Court of Appeal said that contractors had

---

56 *Smith v Eric S Bush* [1989] 2 All ER 514
57 (1986) 6 Con LR 85
58 [1976] 1 All ER 785

a cause of action in negligence against certifiers and valuers. Until *Pacific Associates* it had not been doubted that architects owed a duty to contractors in certifying. In *Arenson v Arenson*,[59] in reference to the possibility of the architect negligently under-certifying, it was said that, in a trade where cash flow is perceived as important, it might have caused the contractor serious damage for which the architect could successfully have been sued. In *F G Minter Ltd v Welsh Health Technical Services Organisation*,[60] the court remarked that an unreasonable delay in ascertainment would completely break the chain of causation, which might give rise to a claim against the architect.

More recent cases,[61] provide firm support to the idea that the reliance principle established in *Hedley Byrne* can be extended to actions as well as advice given by the architect.

33 Is an architect liable for failure to advise a client to use a better material?

An architect's job is to balance many differing factors in giving advice to a client. Such things as compliance with statutory requirements and cost are clearly important, as are durability of materials, maintenance and appearance. Generally speaking, it is always possible to specify a better material even if that means simply thicker (or thinner). Therefore, the question as posed is fairly ambiguous and it is necessary to place it in some kind of context. The question was considered in *Pride Valley Foods Ltd v Hall & Partners*.[62]

If an architect is engaged without the benefit of a proper written contract, it will be contrary to the Codes of Professional Conduct of both the RIBA and ARB. Nevertheless, in law it will be possible to imply the usual duties that an architect owes to his or her client in those circumstances. The same is broadly true of a civil engineer or a quantity surveyor. However, a project manager falls into a different category for which there is no norm in this

---

59 [1975] 3 All ER 901

60 (1979) 11 BLR 1; partially reversed by the Court of Appeal, but not on this point, (1980) 13 BLR 7 CA

61 *Henderson v Merritt Syndicates* [1995] 2 AC 145, (1994) 69 BLR 26; *White v Jones* [1995] 1 All ER 691; *Conway v Crow Kelsey & Partner* (1994) 39 Con LR 1; *J Jarvis & Sons Ltd v Castle Wharf Developments & Others* [2001] 1 Lloyds Rep 308

62 (2000) 16 Const LJ 424

generally accepted sense. That is because the term 'project manager' can encompass a variety of roles.

It was crucial to determine what the project manager had actually undertaken to do. Although this would have been the usual starting point no matter which construction professional was involved, it was especially important where the term 'project manager' did not automatically suggest any particular set of duties.

Hall had been engaged to provide a variety of services, including preparation of the brief for a factory, specification of materials, outline sketch plan and the giving of further advice. After the factory was completed, it was destroyed by fire. The building contained a large number of polystyrene insulation panels and it was said that these substantially assisted the speed at which the fire spread. Pride brought an action against Hall, alleging negligence in failing to warn them of the danger in using the polystyrene.

Importantly, the court found that because Hall had specified the materials, it followed that it had a duty to advise Pride about possible dangers. Although Hall stated that it had advised Pride, the court rejected it on the basis that there was no written evidence of such advice even though there was plenty of written evidence concerning other advice Hall had given. However, the court held that Hall was not liable for the damage, because on the evidence it was clear that, even if Pride had received the advice from Hall, it would have ignored it unless the advice had amounted to a requirement from insurers, planning authorities or the like.

It follows that whether an architect is liable for failure to advise a client to use better materials will depend on all the circumstances. It is one of an architect's tasks to advise a client on all aspects of a proposed project, and part of that will necessarily involve advice about materials. Having said that, it is important to bear in mind that it would be completely unreasonable to expect an architect to give the client information about every single material proposed to be used in a building – still less to advise a client about all the possible alternatives, risks and costs of each material. However, a client has the right to assume that an architect will specify only adequate materials and that, if there is a question mark over any material, the architect will draw it to the client's attention so that a fully informed decision can be made.

Therefore, it is not strictly correct to say that the architect must always advise a client to use a better material. If that were true, architects would be under a duty always to specify the very best material for each case. In practice, there are very many occasions when a material that does not perform as well as other materials will nevertheless be perfectly satisfactory for a particular application. It is part of the architect's skill to know when to use particular materials and when the use of a specific material should be discussed with the client. A lesson to be drawn from the *Pride Valley* case is that all such advice should be confirmed in writing.

34 The job went over time. The employer and the contractor did some kind of deal. Where does that leave the architect?

Employers and contractors are always doing some kind of deal. Often it concerns the final account; sometimes it is about extensions of time or liquidated damages; sometimes it concerns everything. This is particularly common among employers who are in business. They may say that time is more precious than money and, therefore, it suits them to reach a quick and relatively painless agreement even if they have to pay more.

What such employers forget is that they and the contractors concerned have entered into building contracts by which they have agreed that such things as extensions of time, loss and/or expense, liquidated damages and the amount due to the contractor will be decided by the architect. Obviously, it is always possible for the parties to a contract to decide on some change to that contract. If they decide that they prefer to sort out extensions of time rather than let the architect do it, that is a matter for them, but it also affects the architect.

Architects, by their terms of engagement, usually agree to administer the building contract in accordance with its terms. Contrary to the belief of some architects, an architect is only empowered to do that which the contract permits. Therefore, if an employer and contractor decide, for example, that the contract period will be extended by 6 weeks, they are entitled to do so, but that does not mean that the architect has to issue an extension of time for that period. In fact, the architect has no power under the contract to do anything in relation to the agreed extension. If the parties formalise it, the architect can probably regard it thereafter as the date for completion in the contract as varied by the parties. It is that varied date that the architect will have to consider in future if further delays are notified and further extensions may be due.

If the employer and the contractor agree that the contractor should be paid a sum of money as well as having the contract period revised, the architect cannot include the sum in any certificate, because it is not a sum properly due under the terms of the contract, but rather a sum agreed *ad hoc* by the parties. Therefore, it is for the employer to pay the sum directly to the contractor.

There are numerous little agreements and deals into which the parties can enter, leaving the contract terms relatively unscathed. However, certain deals may leave the architect in a difficult position. This can happen if the deals are done without notifying the architect or if the employer begins to interfere with the certification process, for example by insisting that the architect certify some extra agreed amount. The employer may enter into what are often euphemistically called 'acceleration agreements' with the contractor. These are often simply agreements that the contractor will work additional hours and weekends, not that the Works will progress faster.

Many architects finding themselves in this position will wonder where they stand, while desperately trying to administer the contract and adapt its terms to suit the problems that half-considered agreements can throw up. Architects should not do that. If the employer and contractor strike some deal that prevents the architect from properly administering the contract, it is probable that the employer has repudiated the architect's terms of engagement. In these circumstances, the architect should immediately seek expert advice on his or her position. If it is repudiation, it will entitle the architect to accept the repudiation, bringing the architect's obligations to an end and opening the door for a claim for damages against the employer.

35   What are the dangers in a construction professional giving a certificate of satisfaction to the building society?

The first thing to understand is that the certificate required by a building society is an entirely different thing to the certificate that an architect will routinely issue under a standard building contract.

The certificate, or more likely several certificates throughout the progress of the project, is required by the building society to give it insurance against the money it has been asked to lend. Where the builder is registered with the National House-Building Council, the architect will not usually be asked to give a certificate, because the guarantees of the NHBC are usually acceptable to the building society.

If, during or at the end of a project, the employer asks the architect to provide a certificate of satisfaction and if, in the terms of engagement, the architect has not agreed to give such a certificate, there is no obligation to give one. Many architects, and their clients, will argue that if they have been engaged for a full service by their clients, they should be prepared to give the certificate at the end, because to do otherwise is tantamount to saying that they have no confidence in the work they have done. To take that point of view is to misunderstand the purpose and implications of giving such certificates. If an architect has been negligent in performing the services, the client can quite easily take legal action under the terms of the conditions of engagement. Whether such action takes the form of arbitration or legal proceedings through the courts depends on the terms of the engagement. The fact remains that the client has a perfectly adequate remedy for any default on the part of the architect and a separate remedy under the building contract for any defects in construction.

Why does the building society want a certificate before it will lend any money? The building society has an agreement with the architect's client, but it has no agreement with the architect. By completing the certificate, the architect is not only certifying that his or her own work has been performed properly, but also that the contractor's work has been carried out correctly in accordance with the building contract. At one time, architects used to be able to offer their own watered-down version of a certificate, in which their liability was very much restricted.[63] More recently, building societies and banks have insisted on architects signing the building societies' and banks' own forms of certificates.

Sometimes architects do work for developers that are their own builders – in other words, a builder that has decided to build a few 'architect-designed homes' for the speculative market. Usually such a builder could obtain NHBC registration, but it is cheaper to ask the architect to provide a certificate of satisfaction.

The giving of such certificates is very dangerous for the architect, who is thereby exposed to an increase in the number of people who could successfully bring an action for damages, and it

---

63 The Royal Institute of British Architects has devised such a form for use by its members. It can be obtained by contacting the Information Unit.

also significantly increases the liability as noted above. Architects giving such certificates know, or will be presumed to know, that the certificates will be used not only for the purposes of obtaining funding but also for selling the property to future purchasers, who might rely on the certificates in lieu of a building survey. This reliance goes to the very heart of the architect's liability. A building society or a future purchaser cannot bring an action against the architect in contract, because they have no contract with the architect. However, they can bring an action against the architect in tort on the basis of the certificates. Without the certificates, an action against the architect in tort for negligence would be very difficult to sustain.

The basis for the action is an old case called *Hedley Byrne & Co Ltd v Heller & Partners.*[64] Put very simply, if a professional person gives advice to a person or class of person knowing that the advice will be relied on and if the person receiving the advice does rely upon it and, as a result, suffers loss, the professional will be liable for such loss. See Question 32 for the ways in which this liability may be extended beyond professional advice in certain circumstances.

36   Is the client entitled to all the files belonging to construction professionals on completion of the project?

In theory, it depends on whether the professional is acting as principal by providing professional services to a client or whether the professional is acting as agent for a client. In practice these situations sometimes get mixed up. For example, architects often make planning applications or building regulation submissions as agents for their clients. Surveyors may attend meetings on behalf of clients or may be instructed by clients to negotiate boundaries and the like.

However, architects are not acting as agents when they give advice on the suitability of building sites or when they design buildings, administer contracts or act as expert witnesses. Quantity surveyors do not act as agents, but as principals, when they prepare bills of quantities or give cost advice.

If professionals are acting as principals, providing a service to their clients, such clients are entitled to what they have contracted to receive, but no more. For example, a quantity surveyor who has agreed to provide bills of quantities or specifications is obliged to provide such bills and specifications, but not any of the rough calculations or workings that preceded them. An architect must provide designs and working drawings, but not the rough sketches and notes. If a client sends original documents, such as leases or conveyances, they obviously remain the client's property. If they are copies, it depends on circumstances whether they must be returned.

If professionals are acting as agents, their clients are entitled to all the documents created by the professionals or received by them from third parties in the course of discharging their duties.[65] For example, if an architect submits an application for planning permission, the client is entitled to the submission together with the approvals once received.

A professional is entitled to retain all the client's letters and copies of the professional's own letters to the client and all internal memoranda of the practice.[66] It is more likely that papers belong to the client when a professional is acting as agent, and more likely that they belong to the professional when that professional is providing a service.

Obviously, the parties to a contract can expressly agree to vary those general terms. Indeed, many terms of engagement that have been especially drafted for the benefit of the client have clauses that modify the position in favour of the client. Some are modified to such an extent that the client may call for any information required whether or not it is strictly part of the professional's own file.

37 When can job files be destroyed?

This question crops up again and again. Before the days of computer aided design, it used to be a particular worry of architects, who often had huge stocks of crumbling tracing paper, paper prints and slightly more durable linen or plastic film.

The basic answer to the question is that it is impossible to be sure that any particular file will not be required in the future. On that basis, no files should ever be destroyed. It is recognised that such a response is not really helpful for a professional, or anyone else for that

---

65 *Formica Ltd v ECGD* [1995] 1 Lloyd's Rep 692
66 *In re Wheatcroft* (1877) 6 Ch 97

matter, who is anxious to reduce the number of files on the shelf or in storage. The problem is best addressed as a process of elimination.

The normal absolute minimum period for retaining any files is 7 years. That is because 6 years is the time limit for limitation of action where a simple contract is concerned and allowance must be made for the fact that a Claim Form (which used to be called a Writ) can be held for some months before serving. It is also the time limit for holding on to records for company and Inland Revenue purposes. Therefore, if files are to be destroyed before 7 years has expired, there must be a good, positive reason for doing so. There are some obvious examples. It may be that the files relate to a small kitchen extension job which has itself been demolished to make way for another, larger, extension.

Clearly, if a matter is already in litigation or arbitration, on no account must any files be destroyed. Indeed, a person has a duty to preserve all such relevant files and disclose them to the other party if required.

It is also important to understand from when the 7 years runs. It does not run from the date at which the documents were generated, but usually from the last date when the professional had any dealings with that project. In this context, it should be remembered that the period may be restarted or interrupted if the client, however informally, sought additional advice.[67]

That does not mean that as soon as the 7 years has expired it is safe to dump the files; certainly not. For example, many contracts are executed as deeds and the limitation period is 12 years (say 13 years for safety). Under the Latent Damage Act 1986, which amended the Limitation Act 1980, the long stop period for action in negligence was fixed at 15 years. That may seem an exceptionally long period of time, but even that period may be extended if it does not begin to run until certain other criteria are satisfied. For example, it may be dependent on an indemnity. Fortunately, actions in the tort of negligence against professionals are not particularly easy to mount.

In addition, it is always important to examine very carefully the original terms of engagement for a particular project. If they were drafted by the client's solicitors, they may have extended the limitation period under the contract. This is done by crafty wording more often than may be realised.

---

67 *Kaliszewska v John Clague & Partners* (1984) 5 Con LR 62

Generally, it is inadvisable to dispose of any agreements, letters of intent or legal documents such as leases. The safest thing is to keep them for ever.

No documents should be destroyed unless the professional's professional indemnity insurers are happy; perhaps it is going too far to say that they must be happy, but they must not raise any objection.

Some firms have a specific written policy about destroying documents. It can sound impressive when a request for a document is met with the response that 'it is company policy to shred all documents 6 years after their generation'. However, the company will be the loser if a document that could be important in defending the company against legal action has been destroyed. It is no defence at all to quote company policy in these circumstances. The other party will not simply go away.

The question can be answered only on a project-by-project basis, running through a checklist of criteria, including:

- whether it is more than 7 years since there have been any dealings;
- whether a deed is involved that would extend the period to 13 years;
- whether it is likely that any claim in negligence could be brought that would extend the period to at least 15 years;
- whether all fees have been paid;
- whether there is anything about which the professional has always had a nasty feeling
- whether the professional indemnity insurers have raised any objections.

If the files pass all these tests, the professional must try to think if there are any other reasons for keeping the files. If there are, they must be kept.

Finally, the method of disposal must be carefully considered. Some people favour offering the files to the client. This may be perfectly satisfactory if the files are clean and contain only material that the client has already seen. However, if the files contain letters covered in scribbled notes and comments or if the files have documents that the client has not previously seen, it is best not to go down this avenue. The only way to get rid of files properly is to physically destroy them by burning (not very eco-friendly) or by shredding. There are many firms that specialise in such work, but it is essential that the firm is reliable and known. Files cannot simply be taken to the nearest wastepaper collection centre, because they inevitably contain confidential information concerning the professional or the client. News reports occasionally highlight instances where personal records of one kind or another are found blowing about the streets. The fallout of such instances so far as the professional is concerned is disastrous.

6 Fees

38 The consultant's appointment was with a partnership. One partner left and the rest became a limited company. Was the consultant correct to continue invoicing the limited company?

The inability to have a clear picture of the identity of a client gives rise to many problems, often in connection with the recovery or attempted recovery of fees.

A partnership is a group of people who have associated together in business with a view to profit. The formation of a partnership, whether it is a relatively loose arrangement governed simply by the Partnership Act 1890 or whether it is the subject of a detailed deed of partnership prepared by legal advisers, does not create an entity separate to the sum of the individual partners. Partners have joint and several liability, so that a creditor may take legal action against any or all of them. Even if a partnership is dissolved, action can be taken against the individual partners long afterwards, subject only to the Limitation Act 1980.

A limited company is entirely different. When such a company is formed, a legal entity is created that is entirely separate from the members (shareholders) and the directors. The company forms contracts: contracts are not formed by its shareholders. When a director of a limited company enters into a contract, he or she does so on behalf of the company.

If a consultant has an appointment with a partnership, the appointment is not changed just because the partnership or some of the partners decide to reform as a limited company. The consultant's appointment remains with the former partners including, in the instance under consideration, the partner who left the partnership.

Therefore, the consultant is not correct to invoice the limited company. The company has no contract with the consultant and, therefore, it has no liability to pay. The consultant should continue to invoice the partnership or the individual partners.

If the new company wants to transfer the original appointment from the partnership, which seems to make sense in the circumstances, it must be done by way of novation, which has the effect of bringing to an end the original contract between the consultant and the partnership and creating a new contract, on the same terms as before, between the consultant and the new company.

Many consultants change their form of trading from a partnership to either a limited company or a limited partnership without realising that a novation agreement must be executed for each engagement. In such situations, most clients will understand the position, if explained, and they will execute a novation agreement without any problems. Of course, if a client does refuse, the consultant is faced with continuing that particular commission as the original partnership. Simply stopping work on that client's project is not an option. It would amount to a serious breach of contract on the part of the partnership, for which the former partners would each have joint and several liability.

It always pays to take care when executing the original contract. For example, if the other party is a husband and wife, both names should go on the contract.

39   If the tender price has been reduced and the architect has been paid for the reduction work, is the architect entitled to be paid for doing the extra design in the first place?

Unfortunately it is only too common for the lowest tender to be returned in excess of the client's budget figure. This can be due to several causes, only one of which need seriously concern the architect.

If the architect has satisfied the client's brief in terms of design and price, the architect is entitled to the proper fee to the particular stage. It may be argued that, if a tender is returned higher than the figure the client wishes to pay, the architect has not satisfied the brief and, therefore, has no entitlement to the full fee. There are two main reasons why a tender may be high. The first, beloved of architects and quantity surveyors alike but mystifying to clients, is that the general climate of tendering is high. There are variations on this theme, such as that the climate is high in that particular area or that there is a lot of work about and none of the good builders is interested in taking on these Works. None of these explanations or excuses really accounts for the fact that the architect or quantity surveyor ought to be well aware of current trends in tendering. It is rare that the 'general climate' excuse is a good reason for a larger than expected price.

The other main reason is that a mistake has been made in the costing of the job. If the architect did the costing, however informally, that is a problem. The architect should not have a problem if the costing was done by the client's directly appointed quantity surveyor, unless the quantity surveyor can show either that the architect did not provide enough information for the cost estimates, or that the architect did not follow the cost plan.

There are other reasons of course; relatively common is the client who deliberately increases the cost of a project despite being warned of the consequences. The architect should always keep a careful record of such instructions and take the precaution of putting advice about the burgeoning cost in writing.

If a high tender has been returned, but the client has agreed to pay the architect's fees to carry out reductions to the drawings and specification, it suggests that the client is not pointing any finger of blame against the architect. However, if the blame lay with the architect, clearly the client ought not to be required to pay fees to the architect to rectify the error. But in such circumstances, clients may abandon logic and argue that they are not prepared to pay for fees for the design of just those parts of the project that the architect is so assiduously removing so as to lower the cost. Again, if the architect was culpable, the client would be correct and no fees would be payable for the design or the removal of the costly parts. However, it follows that an architect who is not culpable is entitled to both sets of fees, just as an architect who is culpable is entitled to none.

Where an architect culpably fails to design a building to the client's budget the architect cannot claim fees for work aimed at reducing the tenders. Moreover, the architect is not entitled to fees for that part of the original design which is over budget. That may be difficult to determine accurately and the deduction of a percentage is a simple solution.

40 Can the architect claim extra fees for looking at claims?

This question usually arises when the architect has been engaged on the basis of SFA/99 or CE/99 (April 2004 updates), which are in very similar terms. In order to arrive at the answer to this question, it is necessary to look at the terms of the engagement. A client will usually argue that dealing with contractors' claims is part of the architect's normal contract administration duties. However, if the architect is being remunerated on a percentage basis (as is usual), the client will also argue that there is no incentive for the architect to reduce a claim made by the contractor, because the architect's percentage payment will increase as the claim increases. Such an argument says little for the client's opinion of the architect's professional integrity and it is a source of wonder that a client will employ an architect while having such reservations. The latest SFA/99 lays that particular concern to rest, because under 'Definitions' it clearly states that the Construction Cost does not include 'any loss and/or expense payments paid to a contractor ...'.

For a fully designed project the architect's management services include 'Administering the building contract'. Although a number of activities are stated as being included, dealing with the contractor's claims is not one of them. Indeed, the services supplement to Schedule 2 includes 'Ascertainment of contractor's claims' as one of the special services to which the head note refers: 'These Special Services, only required if the need arises during the commission, may be instructed as additional services.' Clause 5.6 deals with 'Other Fees' that are to be calculated on a time basis unless otherwise agreed. It applies especially to situations where the 'Architect is involved in extra work or incurs extra expense for reasons beyond the control of the Architect'. Schedule 2 clearly identifies the ascertainment of claims as additional services and it is to be noted that the phrase 'contractor's claims' is broad enough to cover extensions of time and additional preliminaries in relation to variations as well as loss and/or expense.

Therefore, the architect is entitled to claim additional fees on a time basis for dealing with claims of all kinds provided that the claim has not arisen as a result of a breach of the engagement by the architect. Architects claiming under clause 5.6 are well advised to notify the client as soon as it becomes clear that extra fees will be involved and detailed timesheets should be kept setting out exactly what the architect does in relation to the claim. The timesheets of many architects are totally inadequate as a record of what was actually done.

41   If the architect is engaged at an hourly rate and the client now wants fixed fee to end of job – must the architect agree?

It is surprising how often this problem arises. Many projects start life in an uncertain way and clients often enter into arrangements whereby architects are paid on a time basis. This is usually perfectly satisfactory for the architects concerned because, provided they keep proper and detailed time sheets, they should not lose out since they are being paid for every hour they work.

The situation is quite quickly perceived as unsatisfactory by the majority of clients and phrases such as 'open chequebook' begin to be used. Clients who are quite happy to pay on a time basis while the initial sorting out of the brief and tentative sketch designs are being produced become less happy when the work settles down to fairly routine construction-type drawings and applications to various authorities, which is clearly going to take a considerable time to complete. The answer to issues of this kind depends on the contract of engagement entered into between architect and client, for example SFA/99 or CE/99. It is perfectly possible for such terms to state that RIBA work stages A and B will be on a time basis, but that the remainder will be on a percentage fee or even on a lump sum basis. It is equally possible, but quite rare, for the terms to state that the architect will be paid on an hourly basis for the whole of the services to issuing the final certificate. If the architect is engaged under a binding contract on the basis that the whole of the architectural services will be carried out on an hourly rate, the client has no power to change that arrangement unilaterally. The client may propose a change, but the architect is entitled to refuse. If the client then refuses to pay on the agreed basis or even sacks the architect and engages another, the client probably commits a repudiatory breach that the architect can accept and claim damages.

However, in many cases where the architect is engaged on an hourly rate, the position is often less clear cut. There may be no proper agreement beyond that the architect will perform whatever architectural services are required at a set hourly rate. It is probable that, under such an arrangement, the client is free to renegotiate at any time. Basically, what the client is saying is that if the architect wishes to continue performing architectural services, the architect must agree to a different system of payment. Where there is no binding contract there can be no breach if the architect refuses or if the client decides to seek the services of another architect. Despite the ready availability of RIBA prepared forms of engagement, many architects still carry out work without a proper contract of any kind. This, of course, is contrary to the RIBA Code of Conduct and also to ARB's Code.

42   If the architect was engaged on the basis of SFA/99 and the client is being funded by a public body such as the Lottery, must the architect wait for fees until funding comes through?

Although one is told never to generalise, it is difficult not to do so and, therefore, one might say that architects, as a class, are not noted for their keen business sense, particularly where their own contracts with their clients are concerned. This is something of which many clients are not slow to take advantage.

Architects enter into contracts with their clients in many ways. Although both the RIBA and the ARB Codes of Conduct require architects to consign all their appointments to writing, setting out key points regarding fees, services to be provided and so on, it is still the case that some architects find themselves working for clients on contracts which, if they exist at all, are purely oral. If the clients of these architects drag their heels on payment, there is little to be done without complex legal action.

The remainder of architects either contract on the basis of an exchange of letters, one of the standard RIBA forms or a bespoke set of terms drafted for the client by solicitors – in other words, in writing. That is very important, because agreements in writing fall under the Housing Grants, Construction and Regeneration Act 1996. The Act does not apply to ordinary consumers constructing residential property for themselves, but it is a very useful Act so far as other clients are concerned.

SFA/99 or similar RIBA terms comply with the Act. Therefore, they set out the way in which payment must be made and what is to happen if the client wishes to withhold payment. They also include provision for adjudication – a quick and relatively inexpensive way of settling disputes.

It is quite common for a client to be reliant upon funding from elsewhere. Typically this might be a bank, a mortgage provider, an insurance company or some body such as the Lottery. Perhaps the appointment document includes a term which says that, notwithstanding the payment provisions (which are probably for monthly payment), the architect will be entitled to payment only when funding comes through to enable the client to pay. Sometimes, there is nothing in the terms and the client simply springs the surprise on the architect at their first design meeting or, more likely, not until a fee account becomes seriously overdue. Many architects try their best to accommodate such clients and do not press for payment until funding appears. They do this partly because there is no point in pressing someone who has no money for payment, and partly because architects naturally try to assist their clients, usually far beyond what is required by the law, codes of conduct or sometimes even common sense.

It is surprising that a client will even consider embarking on expensive construction work without having the necessary means to pay and trusting that money will be made available in time. There is no doubt that, under SFA/99, the client is obliged to pay the architect the amount stated on the invoice if no withholding notice has been served.

The architect is not simply entitled to stop work if payment is not made. However, the architect is entitled, under SFA/99 and under section 112 of the Act, to suspend performance of all obligations if payment is overdue provided that at least 7 days' notice in writing is given stating the intention to suspend and the grounds for doing so. Few architects seem to avail themselves of this right, although nothing concentrates a client's mind so well as the knowledge that all work will stop in 7 days. The right is valuable because, contrary to popular belief, there is no such right of suspension under the general law if the client fails to pay, although consistent failure to pay might be grounds for treating the client's defaults as repudiation.

There is another important part of the Act that is not reflected in SFA/99 nor in most other construction contracts. This is a provision that strikes at the heart of the 'pay when paid' ethos. The key part is as follows:

> A provision making payment under a construction contract conditional on the payer receiving payment from a third person is ineffective, unless that third person, or any other person payment by whom is under the contract (directly or indirectly) a condition of payment by that third person, is insolvent.[68]

Therefore, if a client attempts to insert a clause making payment dependent on receiving funds from elsewhere, the clause is ineffective unless the supplier of the funds becomes insolvent. The likelihood of a bank, building society or the Lottery becoming insolvent, while not negligible, is not significant. Therefore, there is no need for the architect to act as the source of a bridging loan until money comes through: it is up to the client to make its own arrangements.

Where the client simply notifies the architect that payment is dependent on a third party after the terms of engagement have been agreed, the architect can, and usually should, simply reject the statement out of hand. There is a danger if the architect gives the client to understand that the architect will wait for the money until it arrives from the funder. In those circumstances, the architect may be estopped (prevented) from subsequently having a change of mind and demanding the money. Estoppel is a legal principle with many facets. In this instance, the principle is that, if the architect represents to the client that strict legal rights will not be enforced and if the client has acted to its detriment in reliance on that representation (such as in this case using its own money which might have been used to settle the architect's debt), the architect is prevented from going back on the representation. The moral is that an architect should insist on being paid on a regular basis. If the architect is moved to allow the client time to pay, that time should have a defined end, which must be clearly set out in writing and acknowledged by the client. Far better for the architect to use the tools set out in the contract and the Act: the right to payment, suspension on failure to pay, adjudication, arbitration or litigation appropriate to secure payment and statutory interest on late payment.[69]

---

68 Housing Grants, Construction and Regeneration Act 1996 section 113(1)
69 Late Payment of Commercial Debts (Interest) Act 1998

43   The architect agreed a fee of 5 per cent of the total construction cost. The contract sum was £325,000, but it is now only £185,000 at final account stage. Is the architect obliged to return some fees?

Many quantity surveyors work on the basis that if they inform the client throughout the contract that the final account is likely to be £x and if, at the end of the project, they are able to tell the client that the final account figure is actually £x minus £20,000, the client will be surprised and grateful and consider the project a huge success. There is a great danger in this approach, which is that the client is anything but grateful and considers that if it had not been for the incompetent reporting of the projected final account throughout the progress of the Works, the client would have known that there was this money to spare, which could have been readily spent on upgrading an important part of the building. The approach smacks of the surgeon informing the patient that death has been avoided when the patient actually would like to be made well.

For architects, this kind of last minute revelation can be catastrophic. That is because it is common for architects to base their percentage fee instalments on the latest valuation or final account forecast by the quantity surveyor. If the quantity surveyor quite suddenly at the end of the project values the final account at something substantially less than forecast, the architect is left with the prospect of having to reimburse some fees.

In this question, the architect is entitled to 5 per cent of £185,000. However, there is a big question to be asked about how the contract sum dropped so markedly. If it was as a result of the client requesting savings, the architect is of course entitled, certainly under the RIBA forms, to recover the cost of redrawing or respecifying, probably at an hourly rate. Where RIBA forms have not been used, the position will depend on the agreement between architect and client, but in most instances the architect should be able to charge for the extra work provided due notice is given to the client before the extra work is undertaken. If the quantity surveyor properly forecast the downward trend in the final account as savings were made, the architect ought to have been aware of the likely outcome and made provision accordingly. Of course, if the quantity surveyor maintained a high forecast until the last minute before dropping it, it may amount to professional negligence.

The answer to the question of what the architect can recover in fees depends on the terms of engagement. If they say 5 per cent of the total construction cost, that is what the architect can charge. If the architect has inadvertently charged more, there must be a reimbursement. If the architect has done extra work, there can be an additional charge, usually on a time basis. There are many pitfalls in claiming fees, often because the parties have not looked far enough ahead or thought of all the possibilities before signing a contract.

Many architects think that to base a percentage fee on the lowest acceptable tender will save the day if the client decides not to proceed at tender stage. Such architects commonly confuse the lowest *acceptable* tender with the lowest tender. If the client does not consider any tender is acceptable, the architect may have a difficult task in arriving at a figure on which to base the percentage fee.

7 Inspection

44 What is the architect's site inspection duty?

'Inspection' and 'supervision' are often confused. Architects are commonly referred to as being responsible for 'design and supervision'. That, of course, is quite wrong. Inspection involves looking and noting, and possibly carrying out tests. Supervision, however, not only covers inspection, but also the issuing of detailed directions regarding the execution of the Works. Supervision can be carried out only by someone with the requisite authority to ensure that the work is undertaken in a particular way. That is the prerogative of the contractor. 'Inspection' is a lesser responsibility than 'supervision'.[70]

Inspection is not something to be carried out lightly. Many architects simply wander on to the site with no very clear idea of what they expect to find, nor indeed what they should be looking for. The architect's appointment document SFA/99 does not deal in detail with inspection of the Works. It simply requires the architect to make visits to the construction Works in connection with general inspection of progress and quality of work, for the approval of any elements reserved for the architect's approval, obtaining information necessary for the issue of notices, certificates and instructions at intervals reasonably expected to be necessary at the date of the appointment and to advise if a clerk of works is necessary. It must be recognised that the number of visits is only an estimate.

It is perhaps cynical to say that a court will find that an architect's duty is to find just those problems that have been missed.

---

70 *Consarc Design Ltd v Hutch Investments Ltd* (1999) 84 Con LR 36

Nevertheless, architects may have difficulty in ensuring that they are not open to legal action from their clients for failure to inspect adequately. The best safeguard for any architect is to be able to demonstrate to a court that their inspection duties were carried out in an organised manner, having regard to what the courts have said. Therefore, before commencing an inspection of the Works, the architect must have a plan of campaign as follows:

• Inspections should have a definite purpose. They should coincide with particular stages in the Works. It is sensible for the architect to sit down beforehand and draw up a list of parts of the construction that must be inspected on that particular visit, together with items of secondary importance to be inspected if possible. The composition of the list and the frequency of inspections will depend on factors such as the employment of a clerk of works, the size and the complexity of the project and the experience and reliability of the contractor. Comments can be made against the checklist as the inspection progresses. The list and the comments are for the architect's own files, not for distribution. Although an architect's inspection duties are quite onerous, he or she will be better able to defend themselves in court against an allegation of negligent inspection if they can show, by reference to contemporary notes, that inspections were carried out in an organised manner.[71]

• Times of inspections should be varied so that a devious contractor cannot rely upon concealing poor work between inspections.

• The architect should always finish an inspection by spending a few minutes inspecting at random.

• Action should be taken immediately the architect returns to the office, whether or not any defects have already been pointed out to the site manager. It is wise to put in writing all comments regarding defective work.

• During site inspections, the architect is bound to be asked to answer queries. It is prudent to give answers on return to the

---

71 *East Ham Corporation v Bernard Sunley & Sons Ltd* [1965] 3 All ER 619; *Sutcliffe v Chippendale and Edmondson* (1971) 18 BLR 149; *Brown & Brown v Gilbert Scott & Payne* (1993) 35 Con LR 120; *Alexander Corfield v David Grant* (1992) 59 BLR 102; *Bowmer & Kirkland v Wilson Bowden* (1996) 80 BLR 131

office when it is possible to sit down and calmly assess the situation. Many decisions made on site are either amended or regretted later.

An architect's failure to inspect will not excuse a contractor from maintaining proper quality control systems. The contractor has undertaken to carry out the Works in accordance with the contract, not carry out the Works to as low a standard as possible unless the architect notices. Nevertheless, although the architect owes no duty to the contractor to find defects,[72] the client is entitled to expect the architect to carry out such inspections as will identify serious defects and a reasonable proportion of minor defects.

If a detail is more complex than usual, the architect will be expected to take more care in inspecting. Just because it is difficult to inspect something does not mean that inspection is not necessary. It is even more necessary, because the contractor might be relying on the difficulty of inspection to attempt to get away with defective work. If work, by its very nature, is being covered up almost as soon as it is done, the architect might argue that there is little point in inspecting because, although the operatives will carry out the work properly while the architect is there, as soon as the architect goes, they will revert to poor workmanship. In fact, that is a cogent reason for continuous inspection of that particular element.[73] The architect's knowledge of the skill and experience of the contractor is an important factor; more time must be spent inspecting the work of an inexperienced contractor.[74] It will usually avail the architect nothing to say that reliance was placed on the contractor's assurance that everything had been properly executed.[75] But if an architect has a great deal of experience of the work of a particular contractor and knows it to be good, reliable and conscientious, less inspection should be needed.

The number of visits and their duration is not the test of adequate inspection. The key is the number of visits necessary. Therefore, it is no defence for an architect to say: 'I visited twice a week for an hour

---

72 *Oldschool v Gleeson Construction Ltd* (1976) 4 BLR 103

73 *George Fischer Holdings Ltd v Multi Design Consultants Ltd and Davis Langdon & Everest* (1998) 61 Con LR 85

74 *Brown & Brown v Gilbert Scott & Payne* (1993) 35 Con LR 120

75 *McKenzie & Another v Potts & Dobson Chapman*, 25 May 1995 unreported

each time.' It may be that in some instances that is unnecessary: in others it is too few or too short. Moreover, an architect should expect to have to visit site more often at some stages of the work, occasionally spending full days if very important work is being done.

The leading authority on the architects' duty to inspect is the decision of the House of Lords in *East Ham Corporation v Bernard Sunley & Sons Ltd*:[76]

> As is well known, the architect is not permanently on the site but appears at intervals, it may be of a week or a fortnight, and he has, of course, to inspect the progress of the work. When he arrives on the site there may be very many important matters with which he has to deal: the work may be getting behind hand through labour troubles; some of the suppliers of materials or the sub-contractors may be lagging; there may be physical trouble on the site itself, such as, for example, finding an unexpected amount of underground water. All these are matters which may call for important decisions by the architect. He may in such circumstances think that he knows the builder sufficiently well and can rely upon him to carry out a good job; that it is more important that he should deal with urgent matters on the site than that he should make a minute inspection on the site to see that the builder is complying with the specification laid down by him ... It by no means follows that, in failing to discover a defect which a reasonable examination should have disclosed, in fact the architect was necessarily thereby in breach of his duty to the building owner so as to be liable in an action for negligence. It may well be that the omission of the architect to find the defects was due to no more than an error of judgment, or was deliberately calculated risk which, in all the circumstances of the case, was reasonable and proper.

These are comforting words, but it is important to give them due weight in the light of the other decisions.

76 [1965] 3 All ER 619

45   Is the architect liable for the clerk of works' mistakes?

The answer to that question depends, in part at least, on whether the clerk of works is employed by the architect or by the client.

In *Kensington & Chelsea & Westminster Area Health Authority v Wettern Composites,*[77] the employer had engaged the clerk of works and was responsible for his payment. The employer took proceedings against the architects, because some of the precast concrete mullions were found to be defective. The court held that the presence of a clerk of works did not remove or reduce the architects' obligation to use reasonable skill and care in inspecting the work, but the employer was vicariously liable for the negligence of the clerk of works although the clerk of works was under the architects' direction. The architects were found liable, but their damages were reduced by 20 per cent to take account of the employer's liability through the clerk of works.

The importance of the clerk of works being employed by the employer was noted in passing in *Gray (Special Trustees of the London Hospital) v T P Bennett & Son.*[78] This was a case where, some 25 years after being built, defects were discovered in the brickwork and supporting concrete nibs of a nurses' home. None of the professionals were found to have been negligent and the defects were found to have been deliberately concealed. In regard to the architect and the clerk of works, the court said:

> ... it is clear that Mr Potts, as clerk of works, was to be the employee of the hospital, even though recommended by the architects. Furthermore, on the evidence it was established that at the end of the job, it was the Bursar of the hospital who notified him that his employment was at an end. I appreciate that the question of control frequently determines who in reality is utilising his services for the purpose of establishing vicarious liability, but in this instance the architect from the outset required an indemnity from the hospital if they were to accept him as an employee. That indemnity the hospital were not prepared to give, with the result ... that the clerk of works remained their employee in the capacity of inspector for the

77   1 Con LR 114
78   (1987) 43 BLR 63

building owner as laid down in ... the contract, although he was also the eyes and ears of the architect.

As an employee of the hospital, the architect was clearly not liable for the clerk of works' actions, albeit in this instance the clerk of works was found to have carried out his duties strictly and no blame whatsoever was attached to him.

8 Defects

46 Can the architect stipulate when the contractor must rectify defective work under SBC or can the contractor simply leave it all until just before practical completion?

During the progress of the Works, the architect is given powers to deal with defects in SBC clause 3.18. There are basically two kinds of defects: those due to an inadequate specification that are not the contractor's problem; and those due to work not being in accordance with the contract. It is only the second kind with which the contract is concerned. The architect may issue instructions regarding the removal from site of any defective work, goods or materials. Nothing in the clause entitles the architect to instruct when the defects must be corrected. This is in accordance with the contractor's right to plan and perform the Works in whatever way it chooses.[79]

If, in the opinion of the architect, the contractor does not comply with an instruction to rectify work not in accordance with the contract, the architect has two possible ways to approach the difficulty. Clause 3.11 gives the architect power to issue a notice to the contractor, giving it 7 days from receipt in which to comply with an instruction. If the contractor fails to comply, the employer may engage another contractor to carry out the instruction and the original contractor will be liable for all the additional costs incurred by the employer, which must be deducted from the contract sum. Such

---

79 *Greater London Council v Cleveland Bridge & Engineering Ltd* (1986) 8 Con LR 30

additional costs will, of course, include any additional professional fees charged to the employer as a result of the contractor's failure. This will be the route of choice in most cases – assuming that a couple of threatening letters do not do the trick first.

As a last resort, the architect may send a default notice to the contractor under clause 8.4.3 giving notice that the employer may terminate the contractor's employment if it refuses or neglects to comply with the architect's instruction to remove defective work and, as a result, the Works are materially affected. This ground used to be qualified by the word 'persistent'. That is no longer the case; the important point is that the Works must be substantially affected. The particular ground appears to be aimed at defects that are about to be covered up or which, for some other reason, would be awkward to put right if not given prompt attention. Therefore, if there is no urgency about the need to make good, this remedy is not appropriate and the contractor is entitled to plan the making good to fit in with its other work.

The position is, therefore, that in principle the contractor is entitled to plan its work, including making good, to suit itself. However, the architect is always entitled to insist on compliance with an instruction within 7 days. In serious cases where the integrity of the Works is threatened, termination can be considered.

47 The contractor incorrectly set out a school building, but it was not discovered until the end of the project when floor tiles in the corridor were being laid. What should be done?

Much depends on the effect of the incorrect setting out. If it resulted in the school encroaching over the boundary on to another person's land, it is virtually certain that, unless a deal can be done with the adjoining owner, the offending part of the school would have to be taken down and rebuilt to a different design. This could be very expensive for the contractor, if indeed the problem was incorrect setting out rather than incorrect setting-out drawings.

There are other possibilities. For example, the school might simply have gained half a metre in length, but it might cause no one a problem. In such circumstances, the school authorities have more school to heat and light, but against that, there is slightly more accommodation. If the gain is minor and of no consequence, it is technically a breach of contract, because that particular part of the Works is not in accordance with the contract, but both parties are likely to let the matter rest. Obviously, the client will not be prepared to pay for the extra walls, floors and roof and they should not be valued.

A trickier difficulty arises if the poor setting out results in the loss of half a metre or the awkward internal arrangement of part of the school. One question the client is sure to ask is why the error was not picked up sooner by the architect. If the error resulted in an internal planning problem it is indeed difficult to see why it was not picked up earlier than when the floor tiles were laid. When the error is picked up only at that late stage, it suggests that it is purely one of length or breadth and it is only when the floor tiling pattern is disturbed that it becomes apparent. It will be for the architect, if challenged, to provide evidence that site inspections were properly carried out and that the average architect in that position would not have found the error.

The basic contract position is that an employer is entitled to get what is being paid for. If I pay for blue boxes, that is what I should have and not green boxes even though the colour may not matter to anyone but me. Alongside that is the rule that where there is a breach of contract the injured party is entitled, so far as money can do it, to be put back in the position it would have had if the contract had been properly performed.[80] However, the courts have modified that rather tough position and they will take all factors into account before agreeing that a contractor in this position must spend large sums of money. The principle the courts apply is that the benefit provided by the remedial work must outweigh the cost of putting it right.[81] In some instances this is easy to calculate. In the *Ruxley* case, a swimming pool was not built deep enough at the shallow end, but the House of Lords decided that it was not worth the cost of demolishing and reconstructing the pool.

In that case, a factor was also whether the injured party would use the money to reconstruct. If something is without value or seriously reduced in value by the error, it is likely that a court would uphold its replacement. On the other hand, it is unlikely that a court would instruct wholesale demolition purely on aesthetic grounds. Therefore, although the baseline is that the contractor is responsible for its errors, care should be taken if the errors are expensive to correct for little apparent benefit.

80 *Robinson v Harman* (1848) 1 Ex 850
81 *Ruxley Electronics and Construction Ltd v Forsyth* [1995] 3 All ER 268

48   What is the position if the employer has taken possession of a block of flats, some have been let and the tenants will not allow defects inspection?

Under SBC, clause 2.38 deals with the inspection of the building for defects at the end of the rectification period. It is the architect's obligation to inspect and deliver a schedule of defects to the contractor no later than 14 days after the end of the period. If the architect fails to do that, the defects still amount to breaches of contract on the part of the contractor, but it cannot be compelled to make them good although it is responsible for the cost of making them good. This cost is measured not on the basis of what it would cost the employer, but what it would have cost the contractor.[82]

So far as the architect is concerned, it matters not whether the flats are let to tenants or whether they are unlet and the employer still has the keys. The architect's obligation to inspect is clearly balanced by the employer's obligation to allow entry to every part of the building. Obviously, any tenancy agreement should contain a clause allowing access to the employer and the employer's agents in circumstances like this. Whether there is such a clause and the tenants involved refuse to comply or whether there is no such clause and the tenants refuse entry is a matter for the employer, not the architect.

An architect in this position should always write to each tenant giving notice of the date and approximate times of inspection and asking to be notified if the date or time is inconvenient. A copy of such correspondence should be sent to the employer. Any tenants who do not allow entry should be sent a further letter and, if that does not work, the architect should obviously speak to the employer and confirm the conversation in writing. If the employer is unsuccessful in securing entry for the architect, there is nothing further the architect can do. The certificate of making good cannot be issued under clause 3.39 and therefore the final certificate cannot be issued. Effectively, the employer has repudiated obligations under the contract and the architect is entitled to accept the repudiation and claim damages (which may be quite slight at this late stage in the project).

---

82 *Pearce & High v John P Baxter & Mrs A Baxter* [1999] BLR 101

There is always the possibility that the employer may instruct the architect not to inspect the flats in question so that a restricted schedule of defects can be issued and the contractor can subsequently receive the certificate of making good. The architect can suggest that course of action, but the architect should not recommend it. In situations like this where the employer is clearly at fault, architects often get into difficulties by failing to properly analyse the position and, instead, they throw themselves into problem-solving mode to attack a difficulty that is not of their making and for the solving of which they are unlikely to receive any thanks. More likely, such architects will eventually be blamed for failing to effect the rectification of defects in the flats in question.

If the architect has no option but to accept the employer's conduct as repudiation, it will be for the employer and the contractor to sort out the rest of the contract. This will be bad news for the employer, because, under clause 3.5, the contractor can expect the employer to appoint a replacement architect. In practice, the situation is likely to be resolved by a commercial deal between the employer and the contractor.

49 The contractor says that, under IC, it has no liability for defects appearing after the end of the rectification period. Is that correct?

The rectification period in all standard building contracts, despite its name, does not signify the maximum period during which the contractor is liable for rectifying defects. It is there for the contractor's benefit. The rectification period in IC is an example. Under the terms of the contract, the contractor's obligation is to construct the building in accordance with the contract documents (clause 1.1), which probably consist of drawings and a specification. If the contractor does not comply with the contract documents, amended if appropriate by architect's instructions, it is in breach of contract.

When the contractor offers the building to the architect as having reached practical completion and the architect has issued a certificate to that effect, the building should have no visible defects and there should be very little work left to complete.[83] The

---

83 *Westminster Corporation v J Jarvis & Sons* (1970) 7 BLR 64

contractor's licence to occupy the site expires at practical completion and it must leave. If there is anything found to be not in accordance with the contract documents and architect's instructions at this point, the contractor is in breach of contract.

If there was no rectification period, the employer would have the right to notify the contractor of the defects, seek competitive quotations from other contractors for making good and then have the defects corrected by the lowest tenderer and recover from the original contractor as damages the total cost of such making good, including professional fees. The employer would have the option to request the contractor to make good the defects at its own cost but, in the absence of a rectification period, the employer would not be bound to do so and the contractor would not be bound to make good although it would be liable for the breaches of contract. The contractor's liability would extend for 6 years from practical completion (12 years if the contract was executed as a deed) in accordance with the Limitation Act 1980.

The rectification period (formerly the 'defects liability period' under previous JCT forms of contract) was introduced to give the contractor the right to return to site and make good any defects notified at the end of the period. It is obviously less costly to the contractor to make good its own defects than to pay the cost involved if other contractors do the work. If the employer does not want the contractor to make good such defects, the architect may issue instructions to that effect to the contractor and an 'appropriate deduction' is to be made from the contract sum (clause 2.10). Unless the reason for the instructions concerns some serious fault on the part of the contractor, such as failure to act despite several reminders, the deduction from the contract sum can be only what it would have cost the contractor to make good.[84]

It is clear from the contract that the contractor's right to return to site extends only to those defects that appear during the rectification period. Any defects that appear afterwards are still breaches of contract and of course the contractor is still liable for them to the end of the limitation period. The employer is entitled to deal with them as though there were no rectification period as noted above.[85]

84 *William Tomkinson and Sons Ltd v The Parochial Church Council of St Michael* (1990) 6 Const LJ 319
85 *Pearce & High v John P Baxter & Mrs A Baxter* [1999] BLR 101

50 Is the employer entitled to hold retention against future defects after the end of the rectification period?

This is an interesting question. Clearly, an employer is not entitled, under any of the standard forms of contract, to keep hold of the retention indefinitely just on the basis that defects might appear. It is probably an attractive idea for the employer, but it amounts to the contractor agreeing to carry out the Works for a sum of money that is less than the tender figure by the amount of the retention percentage.

What lies behind this question is the many occasions when defects are made good by the contractor, but where there may be a doubt about the continuing efficacy of the rectification. Such things as roof leaks or defects in heating systems can seem to have been corrected, but recur with a change in the weather. Indeed, there are some repairs that cannot be verified until the appropriate season comes around again.

GC/Works/1 (1998) gets over the problem in clause 21(4) by providing that, in the case of rectified defects, the maintenance period (as it is misleadingly referred to under this contract) begins again from the date of rectification. JCT contracts do not have this provision, but probably it should be included in the next revision. There is nothing to stop an employer from having a similar clause written into a JCT contract with provision for re-inspection of the rectified work before the certificate of making good can be issued.

Unfortunately, unless the employer has had the foresight, or has been advised, to insert such a clause, the certificate of making good must be issued as soon as all the making good is completed.

It occasionally happens that there is a particularly intractable defect that will not respond to the contractor's frequent attempts to make good. Possibly, even when the contractor finally believes the defect is rectified, there may be a doubt based on more than a possibility. If the architect is not satisfied that the defect has been made good, the certificate of making good must be withheld, thus depriving the contractor of the release of retention under SBC. The architect's powers are circumscribed by the contract of course, but it is open to the architect to suggest to the employer a way out of the dilemma provided the steps are taken before invitations to tender have been sent out. This would amount to a variation to the contract whereby the certificate could be issued, excepting the defects in question, and the retention released except for a suitable sum related to the possibility of dealing with a recurrence of the defect in the future. This will have the advantage of giving the contractor some retention to which it would otherwise not be entitled possibly for a long time, while placing a sum at the disposal of the employer in the event that the contractor later refused to deal with the recurring defect. Much depends on circumstances. To vary the contract as suggested may not be appropriate at all, and it is certainly the case that retaining the whole of the second moiety of retention could have a significant effect on the contractor's attention to the rectification process.

51   The contractor has relaid a defective floor at the end of the rectification period. Can the cost of relaying the carpet be deducted from the final account?

Under clause 2.38 of SBC and similar clauses under other JCT contracts, the contractor is entitled to return to site to make good those defects notified to it in the schedule of defects delivered to the contractor by the architect at the end of the rectification period. The defects in question are defects, shrinkages and other faults that are due to materials or workmanship not being in accordance with the contract or a failure by the contractor to carry out its obligations under the contractor's designed portion. An appropriate deduction is to be made from the contract sum in respect of those defects which the architect, with the employer's consent, has instructed the contractor not to make good.

It is assumed that the carpet, to which the question refers, has been purchased and laid by the employer after practical completion of the Works. The defects are breaches of contract on the part of the contractor. The question is whether the contractor is liable for the cost of having the carpet professionally relaid after the remedial work. The answer to the question depends on the principle of foreseeability. In other words, at the time the contract was executed, was it obvious to the contractor that, if a defect arose in the flooring which had to be put right by the contractor completely relaying the floor, it was likely that the employer would have a carpet laid of the same type and quality as was in fact the case? If the answer to that question is 'Yes', the contractor is liable for the cost of relaying the carpet.[86] However, that cost cannot be deducted from the final account by the architect in the final certificate. The employer has the choice either of taking action against the contractor for the cost, or – the simpler method – after having served the appropriate notices under clauses 4.15.3 and 4.15.4, of setting-off the cost against the amount due in the final certificate.

If it was foreseeable that the employer would lay a carpet, but not of the quality or requiring such care in laying, the contractor would be liable only for the kind of costs that would be reasonably foreseeable.[87]

52 Due to a construction defect in a swimming pool, hundreds of gallons of water have been lost. Can the employer recover the cost from the builder?

This question is essentially the same as the last one. In this instance, it is presumed that practical completion has taken place and the pool filled with water. At some point, cracking occurs and the water escapes. It is conceivable that, if the pool has rooms beneath it, the escaping water may do serious damage to those rooms. For the purposes of this question let us leave aside for the moment the fact that there may be an insurance claim.

We are again looking at consequential losses and the question again is what the contractor had contemplated in the event of a breach of contract. There can be no room for doubt that the contractor knew that the pool would be filled with water and that the consequences of a bad leak would be serious. The contractor is liable to the employer for all these costs and the employer must recover them outside the contract either by taking legal action for recovery or, having issued the relevant notices, by setting-off against a certificate. The damage to the contents of any rooms affected by the leakage would be recoverable by the employer subject to the contractor being aware at the time of executing the contract that the rooms would contain that type of content.

Another possible scenario is that the leak and loss of water occurred during the progress of the Works while the pool was being tested before practical completion. Subject to whatever provisions

---

86 *H W Neville (Sunblest) v William Press & Son* (1982) 20 BLR 78
87 *Hadley v Baxendale* (1854) 9 Ex 341

there might be in the bills of quantities, the contractor would be liable for the cost of the water as before and the contractor would be responsible for dealing with damage to the structure under its ordinary contractual obligations to carry out and complete the Works in accordance with the contract documents.

53   IC: In this project there are two rectification periods: 6 months for building, 12 months for heating and electrical. Should there be two final certificates?

It is very common for an employer to include different lengths of rectification period in the contract. It is usually argued that, although 6 months is perfectly adequate for any defects in the building fabric to come to light, it is important to allow the services to operate throughout the full year to see what effect the variation in seasons may have. That is a reasonable argument so far as the services are concerned, albeit that the seasons during the course of one year may be quite different from the years before or following. However, the same argument can be used for the building fabric. The effect of torrential rain or severe frost or even great heat should be considered.

Unless the employer wishes to use sectional completion, there is no provision for two different rectification periods under IC. Clause 2.30 deals with the rectification period, the length of which is to be inserted in the contract particulars. The term 'Rectification Period' is in the singular. Clause 2.31 refers to the issue of the certificate of making good. Only one such certificate is to be issued. There is no provision for release of any retention at this point. Half the retention is released in the certificate after practical completion and the remainder is not released until the final certificate. Clause 4.14 deals with the timing and issue of the final certificate, of which there is only one. The timing of its issue is dependent upon the later of two events: the sending of the computations of the adjusted contract sum to the contractor or the issue of the certificate of making good.

The final certificate marks the end of the contract. Clearly, therefore, there cannot be two such certificates. Indeed two *final* certificates is not something that it is possible to envisage. Even where sectional completion is validly employed, these are sections of the one contract which is completed by the issue of one final certificate. There can only be two or more final certificates if there are two or more contracts. The question postulates an invalid entry in the contract particulars.

The question is not whether this situation gives rise to two final certificates – clearly it cannot do so – but whether the two different rectification periods are binding on the parties. The answer is by no means clear cut. On the one hand, the employer may say that the contractor's liability to return to site and make good defects, although not its liability for the defects themselves, ends after six months for the building, but the contractor can be recalled to attend to any services defects that arise during the first and the following six months. On the other hand, the contractor may argue that it is entitled to take advantage of the shorter period only, as the two periods were wrongly entered in the contract particulars by the employer. This would result in a quicker release of the retention in the final certificate. On balance, the contractor's argument appears to be the more attractive in principle.

9 Design

54 Can the architect escape liability for defective design by delegating it to a sub-contractor?

In general, the answer to this question is no. However, the precise terms of the architect's engagement by a client will obviously affect that statement. It is possible that an architect can have exclusions of liability for design written into the terms of engagement, but note that such terms will be effective only if they satisfy the test of reasonableness in the Unfair Contract Terms Act 1977. If the architect has entered into an engagement with a client on one of the RIBA terms or on a simple exchange of letters, that architect will have overall responsibility for design of the project. The only way in which the architect can avoid that liability is if the client specifically so agrees.

In *Moresk Cleaners v Hicks*, the client was a dry cleaning and laundry company. It engaged the architect to prepare plans and specifications for the extension of its laundry. Unknown to the client, the architect had delegated the design of part of the building to specialist sub-contractors. Subsequently, cracks appeared in the structure that were found to be design defects. It was held that the architect had no power to delegate duties to others without the permission of the client.[88]

The RIBA terms (SFA/99 and CE/99) contain important provisions that allow the architect to advise on the need to appoint consultants to carry out specialist design (clause 1.6.4). In such instances, the architect is clearly not liable for any defects in the

---

88 *Moresk Cleaners Ltd v Thomas Henwood Hicks* (1966) 4 BLR 50

consultants' designs. This point is emphasised in clause 3.11 where the client undertakes that where work or services are performed by any person other than the architect, the client will hold that person responsible for the competence and performance of the services and for visits to site. This is a most important clause that protects the architect, and a forerunner of this clause has been upheld by the courts for the architect's benefit.[89]

Obviously, if the architect fails to advise a client on the appointment of specialist designers, the client will be entitled to assume that the architect will retain design responsibility in those areas. It is not sufficient that the specification or bills of quantities refer to design by specialist consultants or sub-contractors, nor even that the contract versions incorporating designed portions are used (such as in ICD or MWD). The architect will be responsible for all design unless the client has been expressly informed otherwise and has consented. Clearly this is best done in writing. An architect who does advise the transfer of design responsibility to others has a clear-cut duty to ensure that appropriate contracts and warranties are put in place to protect the client in the event that there are defects in the transferred designs. Failure to put such matters in train may well amount to serious professional negligence on the part of the architect.

55   Does the contractor have any design responsibility for trussed rafters under IC if they are in the specification?

As noted elsewhere, the architect has overall design responsibility, which can be transferred to others only with the express agreement of the employer. The problem with specifying trussed rafters is that the contractor under IC has no design responsibility whatever to the employer. The specification will no doubt say that the roof trusses or trussed rafters are to be supplied by either a sole named supplier or by one of a choice of suppliers. The suppliers in question are actually being asked to design the trusses for the particular building. It may be that the architect has obtained technical information from the suppliers which aims to help the architect choose

---

89 *Investors in Industry Commercial Properties Ltd v South Bedfordshire District Council* (1986) 5 Con LR 1

the appropriate trusses. So the architect may have specified 16 no. type E3 trusses 15m span at 30°. Alternatively, the architect may have left it quite open by specifying 16 no. trusses suitable for 15m span at 30°. Either way, the actual design has been carried out by the supplier.

If there is a failure of the trusses due to faulty design, the contractor may have a remedy against the supplier. This depends upon the terms of the supply contract, but it is likely that there will be a remedy. However, the employer cannot force the contractor to use that remedy, because the contractor has no design liability. If the supplier or the contractor went into liquidation, the employer would have no redress except against the architect. Most contractors in this situation will do everything they can to get the supplier to deal with the failure and the cost of other damage such as broken roof tiles and plaster ceilings caused to the building by the failure. The contractor will be looking to recover the whole of the cost of new trusses and rebuilding the damaged part. However, the bottom line is that, if for any reason the contractor tires of pursuing the supplier, the contractor is entitled to simply turn to the architect and say: 'We did what the specification and drawings said. Part of the building has collapsed. Please instruct us what we should do now.'

Needless to say, the architect would be in a very difficult situation. The answer is for the architect to propose the use of ICD and expressly obtain the employer's consent to the inclusion of the trusses or trussed rafters as part of the contractor's designed portion.

56 How can a contractor be given design responsibility under MW?

In *John Mowlem & Co Ltd v British Insulated Callenders Pension Trust Ltd*,[90] an experienced Official Referee, HH Judge Stabb, said: 'I should require the clearest possible contractual condition before I should feel driven to find a contractor liable for a fault in the design ...'. The judge was considering the JCT 63 form of contract which, like MW, had a clause providing that nothing contained in the bill of quantities should override, modify or affect the conditions of contract except that MW referred to the specification or schedules of work rather than bills of quantities. Like JCT 63, MW requires the contractor to carry out and complete the Works. There is no design requirement in either contract.

It is surprising that, in considering MW 80, which also contained the prohibition against overriding, HH Judge Fox-Andrews, in a subsequent case,[91] thought that a requirement in the specification that the contractor should carry out design did not attempt to override, modify or affect the conditions and he decided that the contractor could be given design responsibility. It does not appear that Judge Fox-Andrews was referred to the decision in *John Mowlem* and it is thought that the decision in *John Mowlem* is to be preferred. Normally, therefore, the contractor will not have any design responsibility under MW and any attempt to pass such responsibility to it by means of a simple insertion in the specification is doomed to failure.

MW, unlike SBC, does not have the facility to make the contractor responsible for the contractor's designed portion although there is such provision if MWD is used. Therefore, something more radical is required. There are probably two solutions.

The first is to require the contractor to enter into a separate design warranty in favour of the employer for the work in question. There is no standard form of warranty for this and the employer would have to have one especially drafted. This should be effective, but it must be understood that it is a separate collateral contract and introduces an element of complication and additional risk in the sense that the two contracts, if challenged, may not necessarily stand or fall together.

The second and probably the better and more straightforward solution is to use MWD, which incorporates a contractor's designed portion. If the contractor's design liability is to be limited to certain elements of the construction, that should be specified in the contract together with the level of design responsibility required.

It is likely that, following *Viking Grain Storage v T H White Installations,*[92] if nothing was stated in the contract, the contractor would have a fitness for purpose liability which is greater than the

---

91 *Haulfryn Estate Co Ltd v Leonard J Multon & Partners & Frontwide Ltd,* 4 April 1990 unreported

92 (1985) 3 Con LR 52

liability of an architect or of a contractor under DB where the contractor's liability is expressly stated to be that of an architect. To avoid any doubt, the liability is expressly stated in MWD.

It is notable that MWD makes no provision for the contractor to take out insurance against any failure in design. The ACA 3 contract does include such insurance provision and indeed it might be worth considering whether the ACA 3 contract should be used instead of MWD if the contractor is to carry out part of the design.

57   Does the architect have a duty to continue checking the design after the building is complete?

A problem that frequently arises is the architect's duty to review the design. In other words, the architect's duty to check whether the design is, in fact, sufficient. Is there such a duty? And if there is, when does it end? If problems arise after the architect has finished any involvement with the building, must the architect return at the request of the client to check the design?

These questions were considered by the court in the case of *New Islington and Hackney Housing Association Limited v Pollard Thomas and Edwards Limited*.[93] Essentially, what the judge was trying to do was decide whether the limitation period had expired so as to protect the architects against an action for breach of contract and/or negligence. In this instance, the judge did indeed find in favour of the architects.

In deciding whether the architect is under a continuing duty to review the design, the starting point has to be the terms of engagement. In this case, the terms included much what one might expect including 'completing detailed design', and the various parts of stage H: tender action to completion. That included preparing the contract, issuing certificates and performing other administrative tasks and accepting the building for the client. The terms did not expressly include a duty to keep the design under review, still less the duty to keep the design under review after practical completion.

The building contract was IFC 84 and the judge was particularly impressed by the fact that, although the contract allowed the architect to issue instructions requiring variations up to practical completion, the contract did not allow the issue of such instructions after practical completion (the same principle applies to the current traditional JCT contracts). So, although the architect could have altered the design before practical completion, the architect was unable to do so afterwards.

The judge went further. He said that if the client asked the architect to investigate a potential design defect after practical completion, the architect was entitled to refuse or to agree only if the client was prepared to pay a fee – because such work was not part of the original terms of engagement. In any event, even if the architect had a duty to review the design after practical completion, it would arise only if something happened that gave the architect reason to think that a reasonably competent architect ought to review the design.

58   Who owns copyright – client or architect?

The straight answer to this question is that copyright in an architect's design is owned by ('vested in' is the legal phrase) the architect who produced the design. Clients sometimes begin to claim copyright if they fall out with their architects and are disinclined to pay the proper fee. Copyright is governed by the Copyright, Designs and Patents Act 1988 as amended and there is also a substantial amount of case law on the topic. It is quite complex and, as with all the other questions in this book, specific problems require specific answers, therefore all that can be done here is to set out a few general principles.

It is important to understand that copyright does not subsist in ideas, but only in the way in which the ideas are presented. Clients often think that they are just as responsible for the finished design as the architect concerned. In a way that is true. Architects and clients usually work together very closely to produce the brief and then to create the building that solves the problem posed by the brief. Some clients have very clear ideas about their requirements, but it is the architect who interprets these ideas in the form of a design. If a client were able to sustain a claim to copyright in a design, it would have to be shown that the client took part in the transforming of the ideas into drawings, whether via drawing board or CAD machine. In the rare case of a client being able to accurately draw a design that satisfied the brief and to pass it to the architect so that all that needed to be done was to draw it out neatly, it might be that the client had a share in the copyright with the architect. However, that will be a very rare circumstance.

Even if the client gives the architect a detailed drawing of what is required, the architect will usually have to change it considerably in order to make it work in practice. It is difficult to show that one design has been copied from another, which is why there are relatively so few successful cases about infringement of copyright. Usually there have to be some significant features on both designs.

In the majority of instances, the architect retains copyright in the designs and the client has a licence to reproduce the design in the form of a building. Sometimes the terms of appointment expressly set this out as in the RIBA-produced forms of appointment (SFA/99, CE/99 and SW/99). Even if the appointment document does not mention copyright, it will be implied that the client has a licence to reproduce the design if a substantial fee has been paid. As a rule of thumb, it is usually assumed that the licence will be implied if the client has paid for all work up to RIBA stage D. If the fee paid is only nominal, no such licence will be implied. For example, a client will not normally have a licence to reproduce a design in the form of a building if the architect has been paid only for preparing a planning application.

10 Possession of the site

59 Can the project manager change the date of possession in the contract?

It is sometimes thought that the architect can issue an instruction to the contractor to change the date of possession in the contract. That view is misguided. The architect can issue only such instructions as are empowered by the terms of the contract (clause 3.10 of SBC), and changing the date for possession is certainly not empowered by any clause in JCT contracts. With the popularity of the project manager among clients, a slightly different view has taken hold that the project manager can change the date of possession. This is consistent with the view that the project manager, simply on the basis of his or her appointment, has wide powers under the contract. This view is, if anything, even more misguided.

The date for possession is one of the most important terms in the contract. The employer's obligation is to give the contractor possession of the site on that date (clause 2.4 of SBC and IC). Failure to give such possession is a serious breach of contract unless the employer has exercised the right to defer possession.[94] The status of a project manager is not easy to define; the powers and duties are not obvious, they depend largely on the discipline of the project manager,[95] but also on the terms of appointment. Usually, a project manager is appointed as the employer's representative, rarely as the contract administrator, and it is even rarer for the project manager to be given all the powers of the employer. Effectively, therefore, the project manager, as normally

---

94 *Freeman & Son v Hensler* (1900) 64 JP 260
95 *Pride Valley Foods Ltd v Hall & Partners* (2000) 16 Const LJ 424

appointed, does not even have power to enter site without the permission of the contractor or the authorisation of the architect.

Even if the project manager was appointed agent with full powers by the employer, the project manager would not have the power to unilaterally change the date for possession; the employer does not have that power. Only the employer and the contractor together may vary the terms of the contract.

60   This is a refurbishment contract for 120 houses under SBC. The bills of quantities say that the contractor can take possession of eight houses at a time, taking possession of another house every time a completed house is handed over. The contractor wants possession of all 120 houses at once. Is the contractor correct?

The answer to this question proceeds from this: there is an implied term in every building contract that the employer will give possession of the site to the contractor within a reasonable time. This means that the contractor must have possession in sufficient time to enable the completion of the Works to be achieved by the contract date for completion. Under the terms of SBC, clause 2.4 stipulates that the contractor must be given possession on the date set out in the contract particulars.

If the employer fails to give possession on the date stated, it is a serious breach of contract. If, as sometimes happens, there is no express term dealing with the topic, a term would be implied and the failure would be a breach of such a term. Failure to give possession is a breach of such a crucial term; if the failure is continued for a substantial period, it may amount to repudiation on the part of the employer. If the contractor accepts such a breach, an action for damages may be started, which would enable the contractor to recover the loss of the profit that would otherwise have been earned.[96] Generally contractors are not anxious to treat the breach as a repudiation, but simply as a breach of contract for which they can claim damages for any loss actually

96 *Wraight Ltd v P H & T Holdings Ltd* (1968) 13 BLR 27

incurred.[97] SBC contains provision in clause 4.23 that allows the contractor to recover such losses through the contract mechanism (clause 4.24.5) and hopefully avoids the difficulties resulting from accepted repudiation. However, it should be noted that the contract provisions do not displace the contractor's right to use common-law remedies if so inclined.

The position envisaged in the question is still quite common, particularly in local authority housing contracts and it was well stated as follows:

> Taken literally the provisions as to the giving of possession must I think mean that unless it is qualified by some other words the obligation of the employer is to give possession of all the houses on 15 October 1973. Having regard to the nature of what was to be done that would not make very good sense, but if that is the plain meaning to be given to the words I must so construe them.[98]

This was a case where the right to possession had been qualified in the appendix to the JCT 63 form of contract. In order to achieve possession in parts under SBC, it is necessary to complete the contract particulars accordingly. Possession as described cannot be achieved by anything in the bills of quantities, because clause 1.3 makes clear that nothing in the bills can override or modify what is in the printed contract. This is a clause peculiar to JCT contracts that still catches out the unwary.

Part of the judgment in *London Borough of Hounslow v Twickenham Garden Developments Ltd*[99] has helped to give rise to the myth that a contractor can be given possession of the site in parts. The court referred to possession, occupation and use as necessary to allow the contractor to carry out the contract. Because this was not something that the court had to decide, the statement does not have binding force.

The idea that a contractor is entitled only to 'sufficient possession' and that, therefore, the employer need give only that degree of

---

97 *London Borough of Hounslow v Twickenham Garden Developments Ltd* (1970) 7 BLR 81

98 *Whittal Builders v Chester-Le-Street District Council* (1985) 11 Con LR 40 (the first case)

99 (1970) 7 BLR 81

possession that is necessary to enable the contractor to carry out work is misconceived. In any event, the contractor is entitled to plan the carrying out of the whole of the Works in any way it pleases.[100]

Although it is not binding authority, the commentary in one of the Building Law Reports sets out the position:

> English standard forms of contract, such as the JCT Form, proceed apparently on the basis that the obligation to give possession of the site is fundamental in the sense that the contractor is to have exclusive possession of the site. It appears that this is the reason why specific provision is made in the JCT Form for the employer to be entitled to bring others on the site to work concurrently with the contractor for otherwise to do so would be a breach of the contract ...[101]

This is an eminently sensible view. Although an earlier JCT form of contract was under consideration, the view is equally valid in the context of SBC. Whether or not the contractor has been given sufficient possession is a matter of fact. In another case, under the JCT 63 form, although the employers were contractually obliged to give the contractor possession of the site, they could not do so. This was due to a man, a woman and their dog occupying the north-east corner of the site by squatting in an old motor car with various packing cases attached and the whole thing protected by a stockade occupying part of the site. Although the precise period was in dispute, it seems to have been about 19 days before the site was cleared and the contractor could actually get possession of the whole site. The court held that the employers were obviously in breach of the obligation to give possession on the contractual date. The contractor could enter on to the site, but it was unable to remove the rubbish and its occupants and, therefore, the breach was the cause of significant disruption to the contractor's programme.[102]

---

100   *Wells v Army & Navy Co-operative Society Ltd* (1902) 86 LT 764
101   *The Queen v Walter Cabot Construction* (1975) 21 BLR 42
102   *Rapid Building Group Ltd v Ealing Family Housing Association Ltd* (1984) 1 Con LR 1

In another case, it was held that the phrase 'possession of the site' meant possession of the whole site and that, in giving possession in parts, the employer was in breach of contract and the contractor was entitled to damages.[103]

The item in the bills of quantities cannot override what is in the printed contract (clause 1.3) and the contractor is correct in requesting possession of all the houses on the date of possession, because that is what the possession clause (2.4) states.

---

103 *Whittal Builders v Chester-Le-Street District Council* (1987) 40 BLR 82 (the second case)

11 Architect's instructions

61 What counts as an instruction?

This is a common question. Standard building contracts refer to instructions and whether they must be in writing or oral, how they may be confirmed and by whom, but contracts do not specify what constitutes an instruction. Usually, to qualify as a written instruction, there must be an unmistakable intention to order something and there must be written evidence to that effect. Not all written instructions are clear – some are decidedly vague (contractors might believe deliberately so). Although an instruction may be implied from what is written down, it is safer from the contractor's point of view to ensure that the words clearly instruct. To take a common example: a drawing sent to a contractor with a compliments slip is not necessarily an instruction to carry out the work shown thereon. It may be simply an invitation to the contractor to carry out the work at no cost to the employer, it may be inviting the contractor's comments, or it may simply be saying: 'This is what we thought about doing, but we changed our minds'. Although most adjudicators would no doubt assume that a drawing sent with nothing but a compliments slip was an instruction to do the work shown on the drawing, such an assumption would be subject to challenge.

The same comment applies to copy letters sent under cover of a compliments slip. Architects sometimes send a letter to the employer saying that they are going to instruct the contractor to do certain additional work in accordance with the employer's wishes. Those same architects misguidedly believe that if they send a copy of that letter to the contractor, it amounts to an instruction to the contractor to get on with the work. Clearly, that is wrong. An instruction on a printed 'Architect's Instruction' form is valid if signed by the architect. An ordinary letter can also be a valid instruction. If the architect wishes, he or she can write the instruction on a piece of old roof tile or on the side of a brick. Providing they are signed and dated and legible, they are all valid instructions. The minutes of a site meeting may be a valid instruction if the contents are expressed clearly and unequivocally and particularly if the architect is responsible for the production of the minutes. However, site meeting minutes are obviously not a good medium for issuing instructions, because of the possible delay in distribution.

62 What can be done if a contractor refuses to carry out an instruction and refuses to allow the employer to send another contractor on to the site?

Clause 3.10 of SBC requires the contractor to comply forthwith (as soon as it reasonably can do so) with architect's instructions that are properly empowered by the contract. If the contractor refuses to do so or simply ignores requests to get on with the instruction, the architect is entitled to issue a written compliance notice under clause 3.11. This notice gives the contractor 7 days from receipt to comply with the instruction. If the contractor still refuses, the employer may employ others to do the work and then an appropriate deduction of all the additional costs may be made from the contract sum. So far so good. The question refers to the hopefully rare instance where a contractor refuses to give access to the site to the other contractor engaged by the employer to carry out the instruction.

In a recent case,[104] the Court of Appeal was faced with an interesting conundrum. Many of the details are unimportant for this purpose, suffice to say that an impasse arose between the employer and the contractor because the contractor, having objected to and refused to carry out an instruction, would not allow the employer to bring another contractor on to the site to do it. The employer sought an injunction to prevent the contractor from refusing access. That is the background. It should be said that both parties must have believed that they had good reason for acting as they did up to that point.

---

104 *Bath & North East Somerset District Council v Mowlem* (2004) 100 Con LR 1

Now the position becomes interesting. Courts are generally reluctant to grant injunctions unless there is a true emergency. They will grant an injunction only if the problem is such that no amount of future damages can sufficiently recompense the injured party after trial. For example, a court may well grant an injunction to prevent someone chopping down a five-hundred-year-old oak tree in a prominent position because, once chopped down, no amount of money could restore the tree. However, a court would be unlikely to grant an injunction to prevent the demolition of an ordinary brick wall, because an award of money will certainly be enough to pay for its rebuilding.

The contractor argued that an injunction should not be granted because the contract contained a liquidated damages clause and, if the contractor was ultimately found to be wrong, the liquidated damages would recompense the employer for the resultant delay. The Court of Appeal disliked this argument. In granting the injunction to the employer, they decided that the contractor was in breach of contract for refusing access in this instance and that liquidated damages was not an agreement between the parties that the contractor could continue its breach of contract. Although liquidated damages was ordinarily the most damages that could be recovered for delay in completion, they did not properly compensate the employer for the loss it would suffer by the continuing breach.

On the basis of this case, it seems that employers can expect to obtain injunctions if contractors refuse access to the site to other contractors who have been lawfully engaged under the terms of the contract.

It is sometimes said that liquidated damages are not only damages due to the employer in the case of a breach on the part of the contractor to complete in time but are also to be regarded as the price payable by the contractor for the option of taking longer to complete. This case shows that such a view is not correct.

63   Should AIs be signed by an individual or the firm?

This question crops up from time to time. It is usually asked by architects fearful that an instruction will be invalid if not signed by the correct person.

The simple answer to the question is that an AI may be signed by any person who is authorised to do so. The architect is the person named in the contract. Only the architect may issue certificates and instructions under the terms of the contract, but that necessarily includes anyone authorised by the architect. The architect should be careful to inform all interested parties of the names of all persons authorised to act on behalf of the architect.

Very often, the name of the architect in the contract will be a firm 'XYZ Architects' or some such name. Therefore, the letter informing all parties of authorised persons, must be signed by 'XYZ Architects'. If the firm is a limited company, the signature of a director will do: if a partnership, it should be one of the partners. If it is a limited liability partnership, it ought to be one of the designated members.

Where AIs, certificates or letters are signed by an authorised person, that person should sign 'for and on behalf of'. This is undoubtedly the best method. It is not sufficient that the letter, etc. is on headed paper. The important thing is that it must be plain that the signatory is not signing on his or her own behalf, but on behalf of the architect, be that company, partnership or sole principal. So it is probably sufficient if the name of the firm is typed where the signature would normally go and the authorised person signs immediately underneath. Sometimes people sign the actual name of the architect. For example, if the named architect is John Smith, one of the authorised persons, say Alice Davis, may sign 'John Smith' provided she initials the signature.[105]

The use of mechanical impressions of signatures are of doubtful validity and should be avoided. A mechanically impressed signature together with the initials or signature of an authorised person is probably valid.

64   Can the architect issue an AI if the client says NO?

Obviously, because the client engages the architect for his or her professional skill, the client is able, at any time, to dispense with the architect's services. If the RIBA terms of engagement are not being used, such termination of the architect's engagement may amount to repudiation on the part of the client and the architect may be able to claim substantial damages. Therefore, the client can stop the architect issuing instructions or performing any other duty under the building contract by simply terminating the engagement.

---

105 *London County Council v Vitamins Ltd* [1955] 2 All ER 229

However, the question is whether the employer can stop the architect issuing an instruction without terminating the engagement. The answer to this question lies in the purpose for which the architect may issue instructions. The architect is empowered to issue instructions by various clauses in the building contract. The architect's terms of engagement with the client may well limit the issue of instructions and require the architect to seek authorisation before issuing any instruction that involves the expenditure of additional money.

The contractor, of course, is not interested in the contents of the architect's engagement. Provided that an instruction issued by the architect is empowered by the terms of the contract, the contractor is secure in carrying it out, because it will be paid for it. If the contractor is in any doubt whether the instruction is empowered it can always ask the architect to name the empowering clause (see SBC clause 3.13). If the instruction that the architect proposes to issue has any monetary or design implications, there is little doubt that the client can instruct the architect not to issue and the architect is obliged to comply. The position is less clear in the case of instructions that simply clarify something on a drawing or in a specification. Such instructions actually do not change the Works at all. In a nutshell and depending upon the precise terms of the conditions of engagement, the client can stop the architect issuing any instruction that results in a variation to the Works. It is unlikely that, short of termination, the client can prevent the architect issuing any other kind of instruction.

65   If the employer gives instructions on site directly to the contractor, must the architect then confirm those instructions in writing?

Many employers seem to find it difficult to stay away from site. It should go without saying that employers should never be allowed to visit site unaccompanied. At best they will get a warped idea of what is happening (for example, all the rooms look too small at foundation stage); at worst they may answer questions from the contractor or give instructions even when no questions are asked. The contractor should be carefully briefed at the pre-start meeting always to refer any queries to the architect and never to ask the employer any questions directly. Despite this, instructions may be given directly to the contractor and the contractor may carry out the instruction without reference to the architect. The architect is not obliged to confirm the instructions in writing.

The first thing to establish is why and in what circumstances the instructions were given. The second is to establish the effect of the instructions on the Works as a whole. It may be that the instruction was given by the employer who told the contractor to check with the architect, or it may be that the employer gave the instruction without really understanding what was being asked. Neither of these circumstances exonerates both employer and contractor from the charge of failing to act in accordance with the contract of course, but life is like that. People fail to act as they should.

If the architect decides that the instruction, although given directly, is simply the kind of instruction that, if the employer had asked the architect to issue, would have been issued without difficulty, the architect will presumably have no problems with ratifying the instruction. The position becomes more difficult if it is an instruction that the architect would not have issued and which perhaps has a detrimental effect on the project. There is no doubt that the employer and contractor, as parties to the contract, are entitled to vary the terms as they wish. If the employer decides to give a direct instruction, albeit the contract provides for only the architect to do that, and if the contractor accepts the instruction, it is likely that either a fresh little contract has been formed for that item of work or, alternatively, it may rank as a variation to the original contract. Obviously, the architect cannot include the value of such a variation in a certificate unless it is the subject of an architect's instruction. If the architect does not confirm with an instruction, the cost of the variation must be paid by the employer directly.

A contractor who accepts a direct instruction from the employer is unwise. If the contractor carries out the work, but the employer contends that the instruction was never given, the contractor is in breach of contract and can be obliged to amend the work to conform to the contract documents.

12 Valuation and payment

66 MW: Can the contractor insist on agreement on price before carrying out the work?

MW deals with instructions in clause 3.4. Variations and the valuation of such instructions is covered in clause 3.6. Clause 3.6.2 states that the architect and the contractor must endeavour to agree a price before the contractor carries out an instruction, but clause 3.6.3 sets out the architect's power to determine the value of work if no agreement can be reached.

Some contractors have had bad experiences with instructions in that they allege that they comply with an instruction and wait a long time for or never receive payment. Where the amounts are relatively small, it is no use arguing that the contractor has the option of seeking adjudication; the cost would be prohibitive.

The contractor can certainly insist that the architect endeavours to reach an agreement first, but it is impossible to insist that another person agrees anything to which he or she objects. Clauses requiring the architect and the contractor to try to agree something are not much use. Can the contractor argue that the architect did not 'endeavour' to agree a price. Yes, of course, but it will be a well nigh impossible task to prove. The architect need only say that they could not agree on the price. It is as simple as that. The contractor cannot refuse to comply with an instruction because there has been a failure to agree the price. It should not be forgotten that the fact that the contractor has carried out additional work does not entitle the contractor to payment. There must be an instruction properly issued under the terms of the contract. If the instruction is properly issued, the contractor should not be out of pocket.

67   Is the contractor obliged to stick to a low rate in the bill of quantities if the amount of work is substantially increased?

Contractors occasionally insert the wrong rate in bills of quantities. Sometimes it is done on purpose. But even if it can be conclusively demonstrated to be inaccurate, it is of no consequence; the rate or price in the bills must be used as the basis for valuation and it can be adjusted only to take account of the changed conditions and/or quantity. The contractor has contracted on the basis that variations may be ordered in the Work, and the employer has contracted to pay for them on this basis. Neither party can avoid the consequences on the grounds that the price in the bills was too low. The contractor's only hope is that it can be shown that the rate in question is narrow in its application and, therefore, not capable of being applied if the amount increases. The matter has been settled by the courts long ago[106] and revisited recently with essentially the same result.[107] A contractor will sometimes take a gamble by putting a high rate on an item of which there is a small quantity or a low rate on an item of which there is a large quantity in the expectation that the quantities of the items will be considerably increased or decreased respectively. If the contractor's gamble succeeds, it will make a nice profit. It is not unlawful, but rather part of a contractor's commercial strategy.[108]

68   SBC With Quantities: The contractor put in a very high rate for an item of which there were only 3 no. in the bills of quantities and it was subsequently found necessary to instruct over 200 no. of these items. Is the quantity surveyor in order to reduce the unit rate?

The answer to this question is virtually the same as the last question. It is the contractor's right to price the bills in any way it chooses. However, the contractor runs the risk that low-priced items may be increased in quantity and high-priced items decreased in quantity. If the contractor is lucky and puts a high price on an item that is subsequently varied so that much more of the item is required, the

---

106 *Dudley Corporation v Parsons & Morrin Ltd*, 8 April 1959 unreported
107 *Henry Boot v Alstom Combined Cycles* [1999] BLR 123
108 *Convent Hospital v Eberlin & Partners* (1988) 14 Con LR 1

contractor gets a windfall. The quantity surveyor is entitled to reduce the unit rate only by a reasonable percentage to reflect economy of scale, but from the starting point of the contractor's bill rate.

69   Can an architect who discovers that the contractor is making 300 per cent profit on some goods it is contracted to supply under MW do anything about it?

An architect might be quite annoyed to discover that a contractor, whose tender has been accepted, is making a large profit on some items. Usually, the architect never gets to know the profit margin because, even where the architect is designated as the person to value variations under MW or MWD, the only relevant document will be the priced specification or a schedule of rates.

Occasionally, the architect does get to know the build up of some of the rates and that is when the nasty surprises occur. Generally, it is unreasonable for the architect to get upset if a contractor is making a large profit. It must be remembered that the contractor has won the contract, presumably, on the basis of the lowest overall tender. Therefore, if the profit margin on some items is high, it is likely to be correspondingly low on others. When contractors submit tenders, they effectively take a gamble. They have to pitch their tenders at a level that will give them a reasonable return, but not so high that they lose the project to another tenderer.

Theoretically, after carefully considering the project and the site, each contractor will look for items that are few in number but which can fairly confidently be expected to increase substantially. They will be given high profit margins on the basis that it will not affect the total price very much, but will eventually net a large profit. On the other hand, numerous items that can be expected to be reduced or even omitted altogether can be priced at a low profit margin or even, occasionally, at a loss, because they will have a big effect on the total price but, if omitted, the possible loss will be omitted also. It is a gamble, because the contractor may be wrong about its expectations. This approach has been accepted as normal practice by the courts.[109]

---

109   *Convent Hospital v Eberlin & Partners* (1988) 14 Con LR 1

The architect can do nothing about the high profit margin if the tender has been accepted. The priced specification is part of the contract and the architect must have regard to it when pricing variations. Architects becoming too enraged at the thought of the 300 per cent profit should consider whether they would want to do something if they discovered that a contractor was making little or no profit at all.

70   If work is being done on a daywork basis, can the time claimed be reduced if the quantity surveyor thinks that the contractor has taken too long?

The whole topic of dayworks is the subject of much misconception. Most standard forms of contract provide for dayworks only as an option to be used if the normal valuation mechanism is not appropriate. It brings up the rear in the valuation tables, because work done on a daywork basis generally costs more than work valued in any other way. Quantity surveyors tend to be frustrated by this state of affairs and use their own experience to reduce the time claimed if it appears to them that it is longer than it should be. It is in those last five words that the misconception lies.

If the parties have agreed that payment is to be made on a daywork basis, the quantity surveyor has no right to reduce the hours and other resources on the sheets.[110] That is because they have agreed that the contractor will be paid for the hours spent and the resources used, not for the hours that should have been spent and the resources that ought to have been used. Of course, the proviso is that daywork is the agreed form of payment. A contractor is not entitled to be paid on a daywork basis simply because it submits daywork sheets. A contractor will often submit such sheets, because usually payment on that basis is better than valuation at contract rates.

Often the magic formula 'For record purposes only' is added. However, where dayworks is to be the method of valuation in any particular case, the addition of those words has little practical value and certainly do not prevent the contents of the sheets being used for calculation of payment.[111]

---

110 *Clusky (trading as Damian Construction) v Chamberlain,* Building Law Monthly, April 1995, p.6
111 *Inserco v Honeywell,* 19 April 1996 unreported

71   The architect and the quantity surveyor cannot keep up with the volume of daywork sheets. Is it OK to sign without checking?

It is never a good idea to sign anything at all without checking. Indeed, it probably amounts to negligence. Therefore, the answer to the question is 'No'. However, that is not the end of the matter. So far as daywork sheets are concerned, the question is what the architect and quantity surveyor should check.

The architect should check that the work has been done and that any equipment mentioned was actually used, together with any materials. The quantity surveyor should check that the arithmetic is correct. As noted above, neither the architect nor the quantity surveyor has the right to check whether the work could have been done in a shorter period of time.

If the volume of daywork sheets is getting too great, it usually indicates one of two things: either the architect and/or the quantity surveyor have let the sheets pile up instead of dealing with them at the right time; or the contractor has not submitted them at the right time. SBC clause 5.7 states that daywork sheets (which it calls 'vouchers') should be submitted to the architect for verification not later than the end of the week following the week in which the work was carried out. At its worst, that could be nearly two weeks – which does seem to be too long. It appears that sheets that are submitted on time but not signed are to be considered as evidence of the work done, unless they can be demonstrated to be inaccurate.[112]

The position is less clear if sheets are submitted late. Late submission cannot, of itself, invalidate a valid daywork sheet. Neither will it excuse the employer from an obligation to pay for the work. If the work is to be paid for on a daywork basis, late submission of the sheets does not remove the obligation, though it may make the evaluation more difficult. It is thought that, in those circumstances, it will be a matter of evidence as to whether or not the work has been done (relatively easy to determine) and how long it took the contractor to do it. In the absence of other evidence, the unsigned daywork sheets will be powerful evidence of the length of time it actually took the contractor to complete the work.

112 *JDM Accord Ltd v Secretary of State for the Environment, Food and Rural Affairs* (2004) 93 Con LR 133

See also the previous question with regard to daywork sheets in general.

72   Is the contractor entitled to loss of profit if work is omitted?

If the contractor has undertaken under a contract to do a certain amount of work for a stated sum of money, it has the right to do it if it is to be done at all. If the contract provides that the work may be omitted, that allows the architect to instruct that the work is to be omitted. However, it does not permit the work to be given to someone else, because that would not be omitting the work but merely transferring it to another party. Architects sometimes wonder if the problem can be overcome by omitting the work from the contract and not giving the work to another contractor until much later in the contract or even after practical completion has been certified. Such an action is likely to be ineffective before practical completion. Whether it would be effective after practical completion is open to question. The key point might well be the intention of the employer at the time the omission was instructed by the architect.

An American case dealt with a contract that is similar to JCT contracts.[113] The contract provided for the omission of work without invalidating the contract and provided that such omissions should be valued and deducted from the contract sum. The American appeal court sensibly held that the word 'omission' meant only work not to be done at all. It did not mean that work could be taken from the contractor and given to another contractor. Two English cases have reached similar conclusions.[114]

The position is very straightforward. If the contractor has contracted to do the work, it has the right to do it, and if the work is given to someone else to do, it is a breach of contract entitling the contractor to damages unless both employer and contractor concurred in the action. The damages are calculated on the principle that the contractor is entitled to be put back in the position, so far as money can do it, as if the contract had been properly performed.

---

113 *Gallagher v Hirsch* (1899) NY 45
114 *Vonlynn Holdings Ltd v Patrick Flaherty Contracts Ltd,* 26 January 1988 unreported; *AMEC Building Contracts Ltd v Cadmus Investments Co Ltd* (1997) 13 Const LJ 50

Where work is omitted to give to another contractor, damages usually amounts to giving the contractor the profit it would have earned had it carried out the work. Of course, it may be that the contractor would not have earned any profit – it may even have made a loss. In these circumstances the contractor may be grateful that the burden of carrying out loss-making work has been removed. It hardly needs saying that the contractor is entitled to loss of profit only if a profit would have been earned.

Because this is damages for a breach of contract and not loss and/or expense, there is no power for the architect under the contract to certify such sum to the contractor and, when it is agreed, it should be paid directly from the employer to the contractor without an architect's certificate.

73   Under what circumstances is the contractor entitled to the costs of acceleration?

A case decided in 2000, defined 'acceleration' like this:

> 'Acceleration' tends to be bandied about as if it were a term of art with a precise technical meaning, but I have found nothing to persuade me that that is the case. The root concept behind the metaphor is no doubt that of increasing speed and therefore, in the context of a construction contract, of finishing earlier. On that basis 'accelerative measures' are steps taken, it is assumed at increased expense, with a view to achieving that end. If the other party is to be charged with that expense, however, that description gives no reason, so far, for such a charge. At least two further questions are relevant to any such issue. The first, implicit in the description itself, is 'earlier than what?'. The second asks by whose decision the relevant steps were taken.
>
> The answer to the first question will characteristically be either 'earlier than the contractual date' or 'earlier than the (delayed) date which will be achieved without the accelerative measures'. In the latter category there may be further questions as to responsibility for the delay and as to whether it confers entitlement to an extension of time. The answer to the second question may clearly be decisive, especially in the common case of contractual provisions for additional payment for variations, but it is closely linked with the first; acceleration not

required to meet a contractor's existing obligations is likely to be the result of an instruction from the employer for which the latter must pay, whereas pressure from the employer to make good delay caused by the contractor's own fault is unlikely to be so construed. [115]

Unless expressly so stated in the building contract, the architect has no powers to instruct the contractor to accelerate work. The contractor's obligation is to complete the work within the time specified or, where no particular contract period is specified, within a reasonable time. The employer cannot insist that the contractor completes earlier than the agreed date in the absence of an express contract term.

No JCT traditional contracts give either the architect or the employer power to order the contractor to accelerate. There is, however, such a power in the ACA 3 form, clause 11.8. The contractor should be able to obtain payment where the architect orders acceleration of the work under a term of the contract or the employer and the contractor agree acceleration.

A contractor will sometimes base its case on the architect's failure to give an extension of time. The contractor will often put more resources into a project than originally envisaged and then attempt to recover the value on the basis that there was no realistic alternative, because the architect failed to make an extension of the contract period. A contractor in this situation contends that, as a direct result of the architect's breach, it was obliged to devote more resources to the project so as to finish by the date for completion, otherwise there was a danger that the employer would levy liquidated damages. This claim tends to be advanced whether or not completion on the due date is actually achieved.

Before this argument can be entertained, the key question is: 'What was the true cause of the acceleration?' The contractor's difficulty is that if the architect wrongfully fails to make an extension of time, either at all or of sufficient length, the contractor's redress under the contract is adjudication or arbitration. If the contractor is entitled to an extension of time, it should simply continue the work, knowing that it will be able to recover its prolongation loss and/or expense, and any liquidated damages deducted, by referring

---

115 *Ascon Contracting Ltd v Alfred McAlpine Construction Isle of Man Ltd* (2000) 16 Const LJ 316

the dispute to adjudication or arbitration. The true cause of the contractor's acceleration is not any breach by the architect, but simply a decision by the contractor to put in more resources. Of course, a contractor in this position may not be entirely confident that adjudication or arbitration will result in reimbursement of money lost. There are few certainties and the liquidated damages may be high. Few would pretend that justice will inevitably be done in adjudication, arbitration or legal proceedings. The contractor may consider that it is less expensive to accelerate rather than face liquidated damages with no guarantee that an extension of time will ultimately be made, even without recovering acceleration costs. It may simply be a commercial decision for the contractor. It is thought that a claim of this kind has little prospect of a successful outcome.

A common situation is where a contractor accelerates without any agreement with the employer or instruction from the architect. The result may be that some time is recovered and an extension of time may be avoided. In this situation, a contractor may argue that, had it not accelerated, there would have been a delay to completion. Using a computer model, it may be demonstrated that the completion date would have been exceeded had the contractor not accelerated. Notwithstanding that, in most such cases the contractor will not find it easy to argue that it was doing other than using best endeavours to reduce delay, and there is no clause in the JCT traditional contracts that could be used to reimburse a contractor in this position.

Although acceleration has been considered in another case, the conclusion was so bizarre as to render it extremely suspect.[116] In this case, the court decided that the contractor was entitled to recover the cost of acceleration if an extension of time was justified, but refused, and the liquidated damages were 'significant'. So far so good, albeit somewhat off the mainstream view. However, the court held that the contractor was entitled not only to the cost of acceleration but also to loss and/or expense for the prolongation that would, but for the acceleration, have taken place. On any view, that amounts to double recovery.

To summarise the position:

---

116 *Motherwell Bridge Construction Ltd v Micafil Vakuumtechnik and Another* (2002) 81 Con LR 44

- There is no clause in traditional JCT contracts and nothing which the general law would imply that gives the architect power to instruct the contractor to accelerate.
- The contractor and the employer can enter into a separate agreement to accelerate, but payment cannot be made under the contract.
- A contractor that accelerates without an agreement from the employer cannot recover the costs of doing so except in wholly exceptional circumstances.

74   What is the effect of agreeing payment 'in full and final settlement'?

It is relatively common for payment to be offered 'in full and final settlement'. Great care must be taken when faced with these words. The law is quite complex and based on what is known as 'accord and satisfaction'. This is defined as: 'The purchase of a release from an obligation whether arising under contract or tort by means of any valuable consideration, not being the actual performance of the obligation itself. The accord is the agreement by which the obligation is discharged. The satisfaction is the consideration which makes the agreement operative.'[117] If there is accord and satisfaction, it acts as a bar to any action.

If a person agrees to accept part payment and to release the other from payment of the balance, this will be valid if the agreement is supported by fresh consideration or if the agreement is a deed that requires no consideration. The key point is that the creditor must accept something different from the legal entitlement.[118] The law does not accept that a debt can be discharged simply by payment of a lesser sum. Therefore, if a party is owed £500 and the debtor offers £200 'in full and final settlement' of the debt, the creditor is entitled to take the £200 and subsequently take action to recover the balance.

The payment would be validly made to settle the debt if it were made in a different way or, perhaps, in a different place. So that, if £500 is owed, payment of £200 worth of grass seed could represent true accord and satisfaction.

117 *British Russian Gazette & Trade Outlook Ltd v Associated Newspapers Ltd* [1933] 2 KB 616
118 *Pinnels Case* (1602) 5 Co Rep 117a

Sometimes a cheque is sent on the basis that payment into the other's bank account will signify acceptance 'in full and final settlement'. If the cheque is simply paid into the account, it is likely that a court would deem that it was accepted on the basis it was paid. It is understandable that a party owed a substantial sum with a cheque in its hand will be keen to recover as much as possible and, therefore, will be anxious to bank the cheque. The answer is to write to the sender noting that the cheque is accepted and will be paid into the bank, not in full and final settlement, but as a part payment of money owing. Then, the cheque should be paid into the account a couple of days later. That allows the sender to stop the cheque if it feels so inclined. Perhaps surprisingly, few cheques appear to be stopped in this situation, the sender preferring to rely on the now useless argument that payment is as indicated by the sender's terms despite the note from the receiving party to the contrary. In *Stour Valley Builders v Stuart*,[119] the builders sent an invoice for work undertaken. The Stuarts disputed the amount and sent a cheque for a lesser sum 'in full and final settlement'. The builders cashed the cheque, but telephoned the Stuarts saying that it was not accepted in full and final settlement. The Court of Appeal held that the cheque was not accepted in full and final settlement and, therefore, the builders were entitled to recover the balance. The court said: 'If the creditor at the very moment of paying in the cheque makes clear that he is not assenting to the condition imposed by the debtor, how can it be said that, objectively, he has accepted the debtor's offer.'

However, the comment applies to a situation where the debt is indisputable. If there is a genuine dispute during which one party says it is owed £500 and the other argues that it owes nothing, an offer in full and final settlement by the alleged debtor of £200 which is accepted by the creditor will not enable the creditor to return later for the balance, because accord was reached in settlement of a dispute and the courts encourage parties to settle their differences by agreement. It is crucial to decide whether there is a genuine dispute or whether one party simply does not want to pay. In these situations, legal advice is always necessary.

75 Under DB, the Employer's Requirements asked for special acoustic windows which the Contractor's Proposals did not include. The contract is signed. Can the employer insist on the special windows at no extra cost?

At the root of this question is the priority of documents. In DB, two situations are envisaged: a discrepancy within the Employer's Requirements and a discrepancy within the Contractor's Proposals. In each case, employer and contractor share the duty of informing the other if either discovers a discrepancy. Under clause 2.14.2, a discrepancy in the Employer's Requirements is dealt with in whatever manner is stated in the Contractor's Proposals or, if not so stated, as suggested by the contractor, which the employer can either accept or reject in favour of its own solution. Either way, it is to be treated as a variation (which is the term for a change). A discrepancy in the Contractor's Proposals is covered by clause 2.14.1. The contractor must suggest an amendment and the employer may choose between the discrepant items or the suggestion at no additional cost.

What happens when there is a discrepancy between the Employer's Requirements and the Contractor's Proposals? In this case, the Employer's Requirements asked for special acoustic windows, but the Contractor's Proposals, by accident or design, does not include them. The contract does not expressly address this problem. Footnote [3] emphasises the importance of removing all discrepancies between the two documents. Unfortunately, discrepancies will occur. The usual way of resolving such matters is on the basis of priority of documents.

It is often mistakenly said that the third recital of the contract covers the position and shows that the Contractor's Proposals take precedence. This recital provides that the employer has examined the Contractor's Proposals and, subject to the conditions, is satisfied that they appear to meet the Employer's Requirements. Whatever else may be said about this recital, the use of the word 'appear' and the fact that it is subject to the conditions is significant. Without these, the employer is satisfied that the Contractor's Proposals meet the Employer's Requirements. The addition of 'appear' makes clear that the satisfaction is simply dealing with surface appearance. One might say 'on the face of things' or, as the lawyers used to say before Latin became unfashionable, *prima facie.* The dictionary defines 'to appear' as 'to give an impression'.[120] It is clearly not intended that, under the contract, the employer or his or her advisers are intended exhaustively to check the Contractor's Proposals to *ensure* that they meet the Employer's Requirements. Had such a thing been intended, it would have been easy for the draftsman to have used clear words to that effect. If the employer requested a five-storey office block in the Requirements, the third recital merely records that the employer believes that is what the Proposals provide. That the statement is made subject to the conditions, very clearly tells the reader that the printed conditions have something important to say about the situation.

The wording strongly points to the intention that the Contractor's Proposals will be drafted to meet the Employer's Requirements. In doing so it is merely confirming the philosophy of the contract as can be discerned from the Recitals as a whole. The Contractor's Proposals should be an indication of how the contractor is to comply with the Employer's Requirements – not an indication of how the contractor wishes to construct the project or allocate risk. The wording of the first and second Recitals reflects this.

However, it is misguided to place such reliance on the third Recital, because the role of the Recitals in interpreting a contract is limited. Where the words in the operative part of a contract are clear, the Recitals do not vary that meaning. It is only when the rest of the contract is ambiguous that one turns to the Recitals for assistance. In this instance the contract is clear, as can be seen below. Therefore, the third cecital has no, or limited, relevance to this particular question.

The contract is clearly written with the intention that the Employer's Requirements prevail in the following way:

- Clause 1.3 provides that nothing in the Employer's Requirements or the Contractor's Proposals can override or modify the printed form.
- Clause 2.2 provides that the Employer's Requirements prevail over the Contractor's Proposals where workmanship or materials are concerned. Clause 2.2 states, in part: 'All materials and goods for the Works shall ... be of the kinds and standards described in the Employer's Requirements, or, if not there

specifically described, in the Contractor's Proposals ...'. From that it is clear that it is only if the Employer's Requirements make no mention of the materials and goods, that the contractor can turn to the Proposals. Clause 2.2.2 is in very similar words in respect of workmanship.

- Under the terms of the contract, the employer cannot issue a change instructing the contractor to vary the Contractor's Proposals. Clause 5.1 provides that a change means a change in the Employer's Requirements. Nor can the employer instruct the expenditure of a provisional sum in the Contractor's Proposals (see clause 5.2.3). If the Contractor's Proposals prevailed over the Employer's Requirements, it would prevent the employer from issuing changes in respect of the discrepant parts of those Contractor's Proposals. That cannot be what the contract intended. Such changes go beyond matters of design and construction and embrace sequence of work and access, etc.

- The intention of the contract is that the Employer's Requirements and the Contractor's Proposals should dovetail together. Where they do not do so, it would be perverse to permit the Proposals to take precedence, because the employer is entitled to assume that the contractor is complying with the Requirements.

The answer to the question is clearly that the employer can insist on the special windows, because the Employer's Requirements prevail over the Contractor's Proposals.

13 Certificates

76 SBC: Is the contractor entitled to suspend work under the Construction Act, if the architect has undercertified?

The right to suspend performance of obligations under the contract is contained in section 112 of the Housing Grants, Construction and Regeneration Act 1996. Section 112(1) states:

> (1) Where a sum due under a construction contract is not paid in full by the final date for payment and no effective notice to withhold payment has been given, the person to whom the sum is due has the right (without prejudice to any other right or remedy) to suspend performance of his obligations under the contract to the party by whom payment ought to have been made ('the party in default').

Further sub-sections proceed to stipulate that at least 7 days' written notice must be given, that the right to suspend comes to an end when payment in full has been made and that the person suspending has, in effect, the right to an extension of any relevant contract period.

Under SBC, the architect is required to issue interim certificates under clause 4.9. Clause 4.14 essentially repeats the substance of section 121. The final date for payment is stipulated by clause 4.13.1 to be 14 days after the date of issue of the architect's certificate. Therefore, it is clear that there can be no final date for payment unless the architect issues a certificate. There is authority to say that the employer may well be liable if the architect does not properly comply with his or her duties under the contract, including the duty to certify at the intervals prescribed in the contract.[121] The employer's liability would depend on the employer knowing that, first, the architect had such a duty and, second, that the architect was in breach of the duty.[122]

However, here we are not considering a situation where the architect fails to certify at all, but where the architect certifies a lesser sum than the contractor thinks is due. Therefore, the architect has not failed to carry out the duty to certify. Clause 4.13.5 obliges the employer to pay the contractor the amount stated on a certificate (obviously, this is subject to the right of set-off and notification in clauses 4.13.3 and 4.13.4). Therefore, if the employer pays the amount on an architect's certificate, even if that certificate is seriously undervalued, the employer cannot be in breach of contract.[123] The contractor's right to suspend arises only if the amount due under the contract, in this instance it is the sum certified, remains unpaid after the final date for payment.

Although the architect's failure to certify the proper amount may be a breach of contract on the part of the employer, depending on whether the employer was aware of any under-certification, it is clearly not something for which the contractor can suspend.

It is worth noting that, although the question is couched in terms of suspending work, both section 112 of the Act and SBC go much further and refer to suspension 'of performance'. In other words, the contractor is entitled to suspend anything at all that the contract requires it to do. The fact that the contract requires the contractor to insure the Works and other matters immediately springs to mind. If the contractor not only suspends work but also suspends all its insurances relating to the Works, the employer will be in a very difficult position.

77 SBC: If the project has taken a long time and has not yet reached practical completion, is the architect correct to release 25 per cent of the retention?

Architects can be their own worst enemies, particularly when they begin to feel sorry for a contractor and let that feeling influence their

121 *Perini Corporation v Commonwealth of Australia* (1969) 12 BLR 82
122 *Penwith District Council v V P Developments Ltd*, 21 May 1999 unreported
123 *Lubenham Fidelities v South Pembrokeshire District Council* (1986) 6 Con LR 85

duties under the contract. Architects do not have unlimited discretion under the contract. An experienced Official Referee once said:

> The occasions when the architect's discretion comes into play are few, even if they number more than the one which gives him a discretion to include in an interim certificate the value of any materials or goods before delivery on site ... The exercise of that discretion is so circumscribed by the terms of that provision of the contract as to emasculate the element of discretion virtually to the point of extinction.[124]

The judge was speaking of the JCT 63 form of contract and even the sole emasculated point of discretion is now gone in SBC. The fact is that the architect's powers are governed solely by the terms of the building contract. An architect who exceeds those powers is acting unlawfully and, depending on the circumstances, possibly negligently.

Clause 4.20.2 sets out the way in which the retention is to be released. The contract assumes that the default retention of 3 per cent will be retained. Clause 4.20.2 states that it may be deducted from so much of the total amount that relates to work which has not reached practical completion. Use of the word 'may' indicates that it is a power that may or may not be exercised by the employer. It is clearly not something for the architect to decide. Clause 4.20.3 permits the employer to deduct half the retention from work that has not been the subject of a certificate of making good.

Therefore, it is clear that the architect is not entitled to release part of the retention before the date set in the contract unless the employer expressly consents. The mere fact that the contractor is taking longer to reach practical completion than was anticipated is hardly grounds for releasing retention early, particularly if the reason for the delay is the contractor's own slow working methods. Even if it is acknowledged that the contractor is entitled to an extension of time for the whole of the overrun period, there are no grounds for early release of retention although, if the contractor makes application for loss and/or expense, it may be able to recover interest on retention held during a period of prolongation. If that is the situation, the employer can authorise early release of the whole or part of the retention. However,

---

124 *Partington & Son (Builders) Ltd v Tameside Metropolitan Borough Council* (1985) 5 Con LR 99

the architect must always remember the purpose of the retention fund, which is a kind of insurance for the employer against having to pay the cost of finishing the project using others, rectifying work, etc.

78   Can an architect issue a negative certificate?

Usually, by a 'negative certificate', what is being referred to is a certificate that shows a negative amount owing from the employer to the contractor – in other words, a certificate indicating that the contractor has already been paid too much. There are three questions that arise from that:

1. Are there any occasions when an architect may issue such a certificate?
2. If yes, is the contractor then obliged to pay the negative amount to the employer?
3. If yes, do the provisions about notices, particularly withholding notices, work in reverse?

If one looks at the JCT Standard Building Contract (SBC) there is nothing that states that the architect may issue a negative interim certificate. On the other hand, there is nothing to say that the architect may not issue one.

Interim certificates are dealt with under clause 4. Clause 4.9.1 states that the architect must issue interim certificates 'stating the amount due to the Contractor ...'. Clause 4.9.2 states that interim certificates are to be issued at the periods stated in the contract particulars and after practical completion 'as and when further amounts are ascertained as payable to the Contractor by the Employer ...'. There seems as first sight to be nothing that entitles the architect to issue a negative interim certificate. Indeed, everything points to certificates stating payments due to the contractor only. The provisions for the issue of the final certificate are the only ones which recognise that there may be a payment due to the employer.

However, clause 4.10, referring to the amount stated as due in an interim certificate, provides that it must be the gross valuation less certain other permitted deductions and the amount previously certified. It is clear, therefore, that if the amount previously certified and the permitted deductions are together more than the gross valuation, any certificate then issued would be showing a negative amount. In practice, this situation can easily arise if a previous certificate is overvalued by more than the total of the work done between the issue of the previous certificate and the new certificate so that the contractor has not carried out work to the value of the overvaluation in the intervening period.

Even in this situation, the standard certification forms issued by RIBA Enterprises are, quite rightly, not worded so as to allow the architect to require payment of the balance by the contractor, and an architect who issues the certificate in the form of a letter is not entitled by the contract to word it in any other way. The inescapable conclusion is that the architect may issue a negative certificate, because that is the result of applying the calculation set out in the contract. However, there is no provision for the architect to certify a payment from the contractor to the employer. This is perfectly sensible and in line with the general intention of the contract. Certification is not to provide the contractor with an exact figure. Its purpose is to provide the contractor with cashflow; sometimes the certificate will be slightly less and sometimes slightly more than the amount of work actually carried out.[125]

In the light of those conclusions, question 3 above does not require an answer. In fact, the provisions regarding notices, particularly with regard to withholding, are not written so as to work in reverse. They expressly refer to notices to be issued by the employer and to the contractor. Of course, the notices are included in the contract as a result of the Housing Grants, Construction and Regeneration Act 1996. Section 110 of the Act refers to the giving of a notice by a party not later than 5 days after 'a payment becomes due from him under the contract ...' It has been noted above that there is no payment due to the employer from the contractor under this contract. Section 111 simply states that a party may not withhold payment unless a withholding notice has been given. Applying that to the contract, the contractor would not be entitled to withhold payment unless an effective withholding notice was served. But the contractor would have no need to issue such a notice unless there was a contractual obligation that the contractor should pay in the first instance.

Therefore, the answer to questions 1, 2 and 3 appear to be Yes, No, and Not applicable, but No in any event.

125 *Sutcliffe v Chippendale & Edmondson* (1971) 18 BLR 149

So far as the formal certificate is concerned, clause 4.15.2 provides that the final certificate must state an amount due to the employer or to the contractor as the case may be. However, provisions for notices in clauses 4.15.3 and 4.15.4 refer only to the employer. Nevertheless, if in the final certificate the contractor was found to owe money to the employer and the contractor wished to withhold some or all of that money, it appears that notices would have to be given under sections 110 and 111 of the Act.

79 Under IC, if the time for issuing a withholding notice has expired, but some serious defects come to light, can the employer set-off the value against the amount certified?

This is the employer's worst nightmare. Under clause 4.8.2, not later than five days after the date of issue of an interim certificate, the employer must give a notice to the contractor stating the amount the employer proposes to pay, to what it relates, and the basis on which it is calculated. Under clause 4.8.3, not later than 5 days before the final date for payment of an interim certificate the employer may give a written notice to the contractor stating any amount or amounts proposed to be withheld and the ground or grounds for the withholding. If the notice is not given, clause 4.8.4 makes clear that the employer must pay the amount stated in the clause 4.8.2 notice. If the employer has failed to give a clause 4.8.2 notice, the amount to be paid is the amount stated as due in the certificate.

The scheme of notices is straightforward. The first notice is the employer's opportunity to tell the contractor that the employer disagrees with the architect's certificate. Provided a proper calculation of the money the employer considers is due is given, that is all the employer need pay. If the employer does not give the first notice, it is assumed that the certified amount is correct and the employer's only chance to avoid paying it is to give the second notice with adequate figures and reasons showing why part or all of it is to be withheld. The deadline for the second notice is 5 days before the final date for payment (which is 14 days from the date of issue of the certificate). If serious defects make their appearance after the deadline, the employer has no option but to pay. If the employer fails to pay, there will be no viable defence if the contractor goes to immediate adjudication. There are similar provisions relating to the final certificate in clause 4.14.

It has been known for an employer, who can put before an adjudicator positive evidence about the existence and value of the defects, to persuade the adjudicator to support the withholding even though it was made without proper notice but, to be frank, that depends upon the appointment of an adjudicator with an inadequate understanding of his or her role.

Realistically, the employer must pay and, if the defects are not corrected by the date of the next valuation, the architect must certify the amount properly due taking the defects into account. This may result in a negative certificate. In most cases, the defects will be made good and the overpayment to the contractor will rectify itself as work proceeds. In rare cases, a dispute may develop and the contractor may become insolvent, leaving the employer in the position of having overpaid. In that situation, the employer may well look to see whether some action is possible against the certifying architect on the basis that if the architect had properly carried out inspection duties, the serious defects would have been discovered earlier. Whether that approach would be successful depends on the circumstances of each individual case.

80   If the employer and the contractor agree the final account, should the architect issue a final certificate in that amount?

All the standard form contracts require certificates to be issued by the person named in the contract as the architect or the contract administrator. A certificate is the formal expression of the architect's professional opinion.[126] In short, it is a very serious document and not something to be issued without careful thought.

It is quite common for the employer and the contractor to effectively 'do a deal' at the end of a project and agree between them the amount the employer will pay to close the contract. Such an agreement is often based on the age-old principle of a figure more than the employer really wants to pay and less than the contractor expects. A settlement is sometimes said to be successful when both parties are dissatisfied with it.

126 *Token Construction Co Ltd v Charlton Estates Ltd* (1973) 1 BLR 48

In the normal course of events, the issue of the final certificate under any of the standard forms will be the culmination of a process that has been continuing from the commencement of work on site. During this time, the contract sum is constantly adjusted to take account of variations and any other matters that the particular contract allows to change the contract sum. After practical completion of the Works, if the contractor wishes to submit any further information to the architect (or to the quantity surveyor if the contract stipulates that the quantity surveyor is to value), there is a specific time within which this may be done, usually 6 months. The quantity surveyor completes the adjustment of the contract sum and, after consultation with the architect, sends this figure to the contractor. Within a contract-stipulated timescale, the architect issues the final certificate. This certifies the amount that is due to the contractor and that the amount has been calculated in accordance with the terms of the contract.

Obviously, where a settlement figure has been agreed between the parties to the contract, it has not been calculated in accordance with the terms of the contract. Therefore, the architect cannot certify that it is the amount which is objectively due to the contractor. It follows that if the parties agree the amount payable from employer to contractor (usually) to settle the contract, the architect cannot issue a certificate to that effect. That is because the issue of the final certificate is a procedure under the contract and the architect has the power only to do that which the contract empowers. Any settlement cannot be a settlement *under* the contract, but merely a settlement *of* the contract. The settlement should be separately recorded and signed by the parties as bringing the contract to an end. It is best done in the form of a deed to avoid any question that there is a lack of consideration. Proper legal advice is required.

Architects who take it upon themselves, or who are persuaded by clients, to issue final certificates for the amount of a settlement face the possibility of future challenges and the real risk that such certificates are invalid. Indeed, architects certifying in these circumstances are probably negligent in issuing certificates that they know to be wrong in the sense that they are not properly calculated in accordance with the contract.

81   If the contractor fails to provide the final account documents within the period specified in the contract after practical completion, what should the architect do?

SBC provides in clause 4.5.1 that the contractor must give the architect or quantity surveyor all the documents necessary for adjustment of the contract sum within 6 months after practical completion. The architect or quantity surveyor then has 3 months in which to ascertain any loss and/or expense and prepare a statement of adjustments to the contract sum. The Intermediate Building Contracts IC and ICD have clauses to similar effect.

Most contractors satisfy the requirement by submitting their own version of the final account, often earlier than practical completion. Many contractors, however, fail to provide all the documents necessary for substantiation. Delay in the issue of the final certificate can often be attributed to delays in the provision of this information. Of course, without substantiation, the quantity surveyor is hampered in completing the account. There have been many instances where the final certificate has been held up for literally years, because the information is not to hand. The Technology and Construction Court has considered this problem and given some useful guidance.[127] The JCT 80 form of contract was being considered but the principle holds good for these contracts also.

The court shone some much-needed light on the position when it pointed out that, if the contract had progressed properly, the information required by the quantity surveyor would have been obtained from the contractor during the progress of the Works. Strictly, the quantity surveyor should be keeping the status of the final account up to date throughout the contract period in accordance with any authorised variations. The effect of that would be that, by the time the certificate of practical completion was issued by the architect, the final account should be just about ready so far as the quantity surveyor was concerned. The purpose of clause 4.5.1 is to give the contractor:

> ... a last opportunity to put its house in order and to ensure that the employer's representatives know of the full extent of

---

127   *Penwith District Council v V P Developments Ltd,* 21 May 1999 unreported

the entitlement to which the contractor considers itself entitled and of the evidence to justify the amount of that entitlement.

The court said that if the contractor failed to take advantage of the opportunity, the architect and the quantity surveyor would have to do the best they could using whatever information the contractor has already provided together with their own knowledge of the project. The court made clear that the architect and the quantity surveyor cannot decline to act and, especially, the architect cannot refuse to issue a final certificate, because that would permit the contractor to control its issue and the contractor cannot be allowed to gain an advantage from its own breach.[128] Moreover, and perhaps more surprisingly, the court held that the provision to the contractor of a copy of the quantity surveyor's version of the final account was not necessary as a precursor to the issue of the final certificate. In the court's view, the final certificate itself would be enough to allow the contractor to decide whether it was satisfied with the amount. The contractor was able to seek adjudication or arbitration if dissatisfied.

The ground rules are now clear. The quantity surveyor should keep the status of the final account up to date throughout the progress of the Works, seeking information from the contractor as required. After practical completion, the contractor has 6 months to submit anything further that may influence the final account. In any event, whether submitted or not, the quantity surveyor should proceed with calculation of the final account after the 6 months expires, and the architect should issue the final certificate strictly in accordance with the contract. That provides in clause 4.15.1 that the final certificate shall be issued no later than 2 months after the last of the following three events: the end of the rectification period; the issue of the certificate of making good; or the date when the architect sends the copies of the final account to the contractor. The court made clear that the 2 months is a maximum period and, whenever the last event occurred, the final certificate could be issued the following day.

128 *Roberts v Bury Commissioners* (1870) LR 5 CP 310

82   Is it permissible to issue a final certificate on an interim certificate form?

At first sight this seems to be the kind of question that someone dreams up on a rainy Friday afternoon. On closer inspection, it is a serious question and just the kind of thing that might trouble an architect who has a final certificate to issue but discovers, too late, that the appropriate pad of certificates has run out and not been replaced. I also remembered that a client of mine used to use an interim certificate form for final certificates on a regular basis.

The answer to this question is really common sense. The important point is whether there is any doubt in the mind of the recipient that it is the final certificate. That is important, because in most forms of contract the final certificate is different in purpose and effect from an interim certificate.

There is nothing to prevent the architect issuing a final certificate in the form of a letter provided that it clearly identifies itself as a final certificate and that it contains everything the contract requires of a final certificate. For example, SBC and IC stipulate what must be contained in the final certificate in clauses 4.15.2 and 4.14.1 respectively. Therefore, if the architect feels there is no option but to issue the final certificate on the interim certificate form, the word 'Interim' must be crossed out and replaced with the word 'Final' in bold capitals. If there is anything else on the certificate that is associated only with an interim certificate, it must be obliterated. It really is much easier for the architect to type out a bespoke version of the final certificate and avoid any possible difficulty.

In *Emson Contractors Ltd v Protea Estates Ltd,*[129] the final certificate was issued on the correct form, but a compliments slip was fixed in such a way that the identification as the final certificate was covered. The result was that the recipient, in the absence of the surveying director, believed it to be an interim certificate and precious days were lost. The court's attitude to mistakes on important documents was demonstrated when it was argued that the certificate referred to 'Emson Construction Ltd' instead of 'Emson Contractors Ltd' and it referred to the date of the contract as '16 October 1984' instead of the correct date of '23 October 1984'.

129 (1988) 4 Const LJ 119

The errors were described by the court as 'technical'. The test applied by the judge was whether Emson was in any material respect misled and he found that no one within Emson had the slightest doubt that the certificate referred to the correct contract. It was a matter of fact that most of the letters regarding the contract emanated from Emson Construction Ltd in any event. So far as the date was concerned, the judge held that although the wrong date could be an important point, the wrong date but in the correct month and year will rarely be so.

83   Is it possible to challenge a final certificate a year after it has been issued?

All architects know that the final certificate issued under SBC and many other JCT contracts is conclusive evidence in any proceedings about various aspects of the contract. One of these aspects is the calculation of the final amount due. That means that, unless one of the parties starts proceedings within 28 days of the issue of the final certificate, the sum due is the sum stated on the certificate and, short of fraud or arithmetical error, it cannot be successfully challenged.

That appeared to be the situation in *Cantrell v Wright & Fuller Ltd*.[130] JCT 80 was the contract in use, but the principle is exactly the same for SBC. Various disputes arose and 13 months after practical completion had been certified, the architect issued a final certificate. Actually, that is not strictly true: the architect issued something described as a certificate for payment under cover of a letter that described it as a final certificate. Cantrell, the employer, objected to the architect and the contractor through its solicitors, but more than 3 years elapsed before the dispute was referred to arbitration. In the circumstances, it was agreed that the arbitrator's first task was to decide if the payment certificate was a valid final certificate. The arbitrator decided it was valid – and that seemed to be that. Except that Cantrell was not satisfied and appealed to the court. At the end of a long and interesting judgment, the court decided that it was not a valid final certificate and, therefore, Cantrell was entitled to challenge the sum stated in it.

In coming to a decision, the court made the following important point: 'When the architect certifies, he is recording for the parties

130 (2003) 91 Con LR 97

his professional, personal and objectively arrived at opinion that the fact situation recorded by the certificate is accurate at the time when the certificate was issued.'

The court usefully considered what would make the final certificate invalid. First it said that it was not invalid simply because it had not been issued on a standard form, provided that what was issued was clearly stated to be the final certificate. The certifying process must have been carried out in accordance with the contract and must be the opinion of the certifier and not that of some other person. Second, the timing of the issue of the certificate was not critical so long as the timescale was reasonable.

However, there were a number of other matters that were more important, namely, those matters which were 'conditions precedent' to the issue of a final certificate. A condition precedent refers to something that must happen before something else can take place. If defects had been notified to the contractor after the end of the defects liability period, the final certificate could not be issued until the certificate of making good defects had been issued. The production by the contractor of final account documentation and the quantity surveyor's adjustment of the contract sum were not conditions precedent and the architect could issue the final certificate without them. This certificate did not fall foul of any of these matters.

The court identified four reasons why the document issued by the architect was not a valid final certificate.

- Unfortunately, although practical completion had been certified after the contractual date for completion, the architect had neither issued a certificate of non-completion nor carried out a review of extension of time as specified in clause 25.3.3. This was fatal to the validity of the final certificate.

- There were nominated sub-contractors and an interim certificate, releasing the balance due to them, had not been issued 28 days before the final certificate as required under the contract. There is no longer any provision for nominated sub-contractors under SBC and, therefore, this objection would not now apply.

- No decision had been taken by the architect about sanctioning variations carried out or adjusting the contract sum as a result of acceptance by the employer of defective work. Therefore, the final adjustment of the contract sum had not taken place.

- The time of issue of the final certificate did not comply with the agreed timetable for issue, nor did it satisfy the requirement of reasonableness, failing to allow sufficient time for the employer to consider the final account.

Therefore, the final certificate can certainly be challenged, but the challenge is essentially that the certificate is not validly issued. It is still necessary to formally challenge a valid final certificate by serving the relevant notice in adjudication, arbitration or litigation as the case may be.

84 Under SBC, should the architect issue a final certificate if further defects have appeared in the Works?

This question clearly refers to the situation where the contractor has made good all the defects included in the schedule of defects and a certificate of making good has been issued. Before the architect has time to issue the final certificate, a defect or defects have appeared. These are latent defects; that is, they were not apparent when the architect carried out the inspection at the end of the rectification period. In these circumstances, most architects would be reluctant to issue the final certificate.

Clause 4.15.1 is clear. The final certificate is to be issued no later than 2 months after the latest of:

- the end of the rectification period or, in the case of sections, the last rectification period
- the date of the last certificate of making good (if there is more than one)
- the date when the architect sends copies of the final account to the contractor.

There is no room for manoeuvre here. The issue of the final certificate has a mandatory timescale.

There is actually no cause for alarm. The conclusive effect of the final certificate is set out in clause 1.10. It is no longer conclusive that qualities and standards of materials, goods or workmanship is to the reasonable satisfaction of the architect unless the architect has been ill-advised enough to express reserve about anything being to his or her satisfaction by a note in the specification, bills of quantities, drawings or an architect's instruction. In any event, the final sentence of clause 1.10.1 states that the final certificate is not conclusive evidence that any such materials, goods and workmanship so expressly mentioned or any other materials, goods or workmanship 'comply with any other requirements or term' of the contract.

It follows that the architect is obliged to issue the final certificate, but that it does not affect the contractor's liability for the further defects. Indeed, the employer has the option of issuing the relevant notices under clauses 4.15.3 and 4.15.4 before withholding payment of a sufficient sum to deal with the defects.

85   Under SBC, when is a certificate issued?

Clause 2.30 refers to the 'issue' of the certificate of practical completion. It is to be issued forthwith. Clause 2.31 refers to the architect issuing a certificate of non-completion, a pre-requisite for the deduction of liquidated damages. Clause 4.9 deals with the issue of interim certificates and clause 4.15 governs the issue of the final certificate. Various deadlines relate to the date of issue and it is important to understand what is involved in the issue of a certificate.

The dictionary definition of 'issue' is to send forth, to emit or to put into circulation. The question concerns when that is done. In *London Borough of Camden v Thomas McInerney & Sons Ltd*,[131] the court said: 'In the ordinary meaning of the word "issue", it seems to me plain that something more is needed for a certificate to be issued under clause 3(8) than the mere signature of the architect upon it ...'.

That was a case on the JCT 63 form of contract, but the principle is the same under current forms. The date of issue must be something other than the date on which it is received, because the contract itself makes that distinction. A notice is clearly not 'given' unless the person for whom it is intended has received it. In another case[132] dealing with certification it was said: '... I have some difficulty in thinking that there would be a sufficient compliance ... if the architect certified in writing and then locked the document away and told no one about it.'

131 (1986) 9 Con LR 99
132 *Token Construction Co Ltd v Charlton Estates Ltd* (1973) 1 BLR 48

It seems clear from this that the date of issue is the date the certificate leaves the possession of the architect. Pre-printed certificate forms have a space for the date of issue. It is obviously important that the date inserted is the date on which the certificate is posted or handed to the employer and contractor.

86 Should an architect, working for a contractor under DB, issue certificates?

Architects find themselves working for contractors under DB in two situations. In the first, the architect initially has been engaged by the employer to prepare the Employer's Requirements and possibly oversee tendering, and then moves on by deed of novation to work exclusively for the contractor. Sadly, this is very common although it is obvious that such an architect is in a conflict situation. The only thing that prevents the architect being in breach of the professional Code is that both employer and contractor know and acquiesce in the arrangement. However, that is not enough to ensure that the architect is not placed in some very difficult situations. The sooner architects stop agreeing to novation agreements for design and build the better.

The second situation is where the contractor engages an architect who has not previously worked for the employer.

A certificate is the formal expression of the architect's opinion. Traditional JCT and some other contracts provide for the architect to issue such certificates. One of the most important of such certificates is the interim certificate for payment. DB does not provide for the issue of certificates by an architect or any other person. There is no architect mentioned under DB. The employer's agent may well be an architect, but all that can be issued are notices and statements of various kinds. None of these documents carries the same weight as a certificate.[133] Therefore, not only should the architect not issue certificates when working for a contractor under DB, such an architect also has no power to do so whether working for the contractor or even for the employer.

---

133 *J J Finnegan Ltd v Ford Sellar Morris Developments Ltd* (No. 1) (1991) 25 Con LR 89

14 Sub-contracts

87 Must the architect approve the sub-contractor's 'shop drawings'?

It is not unusual for a contractor to submit a sub-contractor's or supplier's 'shop drawings' for approval before manufacture of the element concerned. Indeed, few sensible contractors would authorise proceeding with manufacture until the architect is satisfied with the details. Of course, in most cases the shop drawings are simply the sub-contractor's own translation of the architect's drawings and details into something that the sub-contractor believes is easier to understand in the context of the particular manufacturing process. In other words, the sub-contractor is using the information provided by the architect through the contractor to produce the shop drawings.

I once knew a very brave architect who would respond to the contractor with the following words: 'If the shop drawings are in accordance with the drawings I have provided, they are correct; if not, they are wrong.' This is equivalent to saying 'check them yourself'. It also requires a large degree of confidence on the part of the architect that the original drawings are completely accurate.

Few architects can say that their drawings are guaranteed to be 100 per cent correct. That is not to criticise architects; it is just a characteristic of the complex nature of the profession that discrepancies and other types of error do occur. Therefore, most architects will check shop drawings just to be sure that their own drawings are correct. The problem is that, in checking whether the shop drawings accurately represent their drawings, architects inevitably check things that have been introduced by sub-contractors. Sometimes, sub-contractors will actually change architects' details to make them suit the particular sub-contract element. Such changes can easily be missed if the architect gives the drawings only a cursory inspection. Architects should either check shop drawings thoroughly or not at all. Even if the architect has no contractual responsibility for checking such drawings, responsibility may be assumed if the architect nonetheless does check them.

In most cases, the architect will want to be satisfied that the shop drawings are accurate and, therefore, will check them. Whether the architect has an obligation to approve the drawings will depend upon the terms of the contract. Such an obligation will usually be found, if at all, in the preliminaries section of the bills of quantities or specification. Ideally, the architect should make sure, before the documents are sent out for tender, that there is no requirement for the approval of the architect. The absence of the requirement for approval will not prevent the contractor from sending the drawings for approval, but it will enable the architect to point out that there is no contractual requirement for the architect's approval. Moreover, the architect should inform the contractor that it is the contractor's task to check and co-ordinate sub-contractors' drawings.

Obviously, if the sub-contractor is being asked to carry out part of the design, the position is rather different. The architect, who is usually the design leader, will have a duty to co-ordinate the sub-contractor's design with the rest. Therefore, the architect will have a corresponding duty to check the drawings to ensure this co-ordination.

The position is, therefore, clear. The architect will rarely have any obligation to approve a sub-contractor's shop drawing unless either the sub-contractor has a design obligation or the contract documents expressly require the architect to approve such drawings. When dealing with the sub-contractor's design, it is safest if the architect avoids using the word 'approve' and instead simply states that he or she has no comment to make. Use of the word 'approve' has been discussed elsewhere.

88 MW: If the contractor is in financial trouble, can the employer pay the sub-contractors directly?

Under JCT 98, there used to be provision for the employer to pay nominated sub-contractors directly in certain circumstances. There are no such provisions in SBC; indeed, there are no nominated sub-contractors in SBC. Even under JCT 98 terms, the direct payment provisions were hedged around by substantial conditions.

Under current traditional JCT contracts, there are no circumstances where the employer should pay sub-contractors directly.

It is important to understand that the employer is in contract with the contractor and the contractor is in contract with the sub-contractors. There is no contractual relationship between the sub-contractors and the employer unless some kind of direct warranty has been employed, because clauses in the main and in the sub-contracts prevent the Contracts (Rights of Third Parties) Act 1999 having any effect. Therefore, the position is that the contractor has undertaken to the employer, for payment, to carry out certain Works. Part of these Works has been sub-let to sub-contractors. That is to say, the sub-contractors have each undertaken to the contractor to carry out their parts of the main contract Works in return for payment from the contractor.

When a sub-contractor carries out work for the contractor, it is part of the main contract Works and the contractor is entitled to payment for it from the employer. If the contractor does not pay the sub-contractor, the sub-contractor's redress is against the contractor. The sub-contractor has no valid claim directly against the employer. It is unfortunate for the sub-contractor (indeed for all concerned) if the contractor gets into financial difficulties or even goes into liquidation. That is the kind of thing that the sub-contractor, like any other business, must try to guard against. Some employers believe that they are entitled to pay the sub-contractor directly and then deduct the money paid from the contractor. Nothing could be further from the truth.

The employer who pays directly will be in breach of the insolvency rules by making the sub-contractor into a preferential creditor. Even if that is not an issue, in the case of a contractor who simply will not pay sub-contractors, the employer will undoubtedly be called upon to pay the contractor in any event for the work carried out as part of the main contract by the sub-contractor. The employer will have no defence. It is not an argument for the employer to say: 'I will not pay you because you have not paid your sub-contractors.' The employer's duty to pay the contractor under the main contract is not dependent on whether the contractor has paid the sub-contractors. Indeed, the contractor's relationship with sub-contractors is no business of the employer's except to the extent that the main contract requires the contractor to include certain provisions in the sub-contract (for example, SBC clause 3.9.2).

On a purely practical level, there is no way in which the employer can be sure that the sub-contractor has not been paid unless the sub-contractor takes legal action against the contractor. In that case, the sub-contractor will recover whatever the adjudicator, arbitrator or judge believes is appropriate.

15 Extensions of time

89 Can the contractor be entitled to an extension of time if it finishes before the date for completion?

The purpose of the extension of time clause in standard form contracts is to give the power to an architect or other contract administrator to extend time if the project is being delayed due to some action or default of the employer or one of the employer's agents. That is so there will always be a proper date for completion from which liquidated damages can be calculated. If there were no such power, time would become at large if, for example, the architect was late in supplying a drawing and the completion date was delayed. The contractor's obligation then would be to complete the Works within a reasonable time.

Therefore, it is clear that the purpose of the extension of time clause is to extend time – nothing more. One of the preconditions before an extension of time can be given is that the completion of the Works is likely to be delayed beyond the contractual completion date (for example, see SBC clause 2.28.1.2). If a contractor finishes before the contractual date for completion, there are no grounds under which an extension of time can be given. Indeed, the contractor has no need for an extension of time, having finished within the contract period.

On one level, therefore, the contractor's request is incomprehensible. However, the reason for the request probably stems from at least one and probably two misconceptions. First, if it is assumed that the contractor put more resources into the project in order to complete on time, the contractor may be wondering how best to recover the cost of the extra resources. The contractor's mind is probably working like this: 'If I had not put any extra resources on to this project, I would have finished late and I may have been able to get an extension of time for that. Therefore, I will ask the architect for an extension of time on the basis that if I had not put on the extra resources, this is when I would have finished. Having established when I would have finished, I can then calculate the equivalent in loss and/or expense for the notional prolongation. That will be what I have saved the employer by putting on the extra resources. Therefore, I am entitled to recover the cost of the extra resources as the cost of mitigation.' That may be what the contractor is thinking, but it is misguided.

The first flaw in the argument is that extension of time has no connection to loss and/or expense. Therefore, even if an extension of time had been possible, it would not lead to recovery of any direct loss and/or expense. The contractor would be better engaged in simply calculating the notional loss and/or expense based on the loss and/or expense clause in the contract.

Second, there is no express provision for a claim for mitigation of costs under most standard form contracts. If the contractor unilaterally decides to put more resources into a project so as to complete it before the completion date, the architect will usually consider that the contractor is simply doing what the contract says and is using best endeavours to avoid or reduce delay. Although this requirement cannot require the contractor to spend more money on the project, there is no mechanism to reimburse a contractor who decides to do so in any event.

If the contractor wanted to recover money, the way forward would have been to have resisted the temptation to put on more resources, but to have notified delays at the appropriate times and made application for loss and/or expense under the correct contractual clause. A contractor who acts first, without proper consideration, is likely to lose out.

90 Under SBC, if the architect gives an instruction after the date the contractor should have finished, is the contractor entitled to an extension of time, and if so how much?

It is often argued, with some merit, that the architect should issue all instructions before the completion date in the contract. But what is the situation if the completion date has passed and the contractor has been trying to finish the Works for several weeks when an instruction for additional work is issued? Clearly, the instruction is going to cause a delay and the contractor is entitled to an extension of time as a result. The question is whether the contractor is entitled to an extension of time from the completion date to the date on which the contractor actually finishes the additional work or is it entitled merely to an extension of time added to the last completion date of a length to represent the time taken to comply with the instruction.

The contractor will usually be looking for the former, but the architect will be anxious to give the latter. The contractor's argument makes sense. It amounts to this: the architect could have issued the instruction at any time up to the date of completion, but chose to issue it after the completion date was past. There is no excuse for that, because where SBC is used, all the work to be done should be known when the Works commence, therefore the architect cannot say that he or she could not issue the instruction until the contractor had reached a particular stage.

The architect's argument is simply that instructions may be issued at any time up to practical completion, therefore although an extension of time is due, it can deal only with the actual period of delay. The contractor is effectively saying, 'How can you give me an instruction to carry out work today knowing that it will take until next week to do, but at the same time you are giving me an extension of time that says I should have finished several weeks ago?'

The question was considered by the court in *Balfour Beatty Ltd v Chestermount Properties Ltd* [134] in connection with the JCT 80 form of contract. The court decided that the contractor was entitled only to the net amount of the delay added on to the completion date. This did not seem to be a result of a strict reading of the contract but rather the court's view of what was a reasonable solution to the problem. The decision, therefore, leaves some unanswered questions – principally what would have been the result if the decision had been appealed? Nevertheless, the decision represents the law on this point until such time as the question is brought to the Court of Appeal.

134 (1993) 62 BLR 1

91 If, under SBC, the architect does not receive all the delay information required until a week before the date for completion, must the extension of time still be given before the completion date?

This question originates from the fact that clause 2.28.2 states that the architect must notify the contractor of the decision in writing 'as soon as is reasonably practicable' but it must be within 12 weeks of receipt of the required particulars. The clause goes on to state that, if the period between receipt and the completion date is less than 12 weeks, the architect must endeavour to notify the contractor before the completion date. This wording, and similar wording in the predecessor contract JCT 98, is a source of concern to many architects. The current wording stipulates that the architect must 'endeavour' to notify before the completion date. That is new wording introduced with SBC and requires the architect to make an earnest or determined attempt. The previous wording let the notification depend on whether it was 'reasonably practicable' to do so.

The key is the date of receipt of the required particulars. The particulars are detailed in clause 2.27.2 and are said to include the expected effects of the relevant event and an estimate of any expected delay. It is only when the architect receives this information that the time period for making a decision begins. Therefore, if the architect does not receive the particulars until a week before the completion date, the architect is still expected to 'endeavour' to reach a decision and notify the contractor before the completion date; but realistically it is probably unlikely.

Contractors sometimes delay providing the information until there are only days to go before the completion date in the hope of being able to argue that the architect has failed to act and, therefore, time is at large. Such arguments have little chance of success. Indeed, if the architect cannot reach a decision before the completion date, despite endeavouring to do so, the matter is then part of the decision to be made by the architect under clause 2.29.5. This clause states that after the completion date the architect *may* and not later than 12 weeks after the date of practical completion the architect *shall* (must) carry out a review of all extensions and make what is effectively a final decision. Therefore, if the contractor delays in providing the required particulars, it runs the risk that a decision will not be made in respect of those delays until some weeks after practical completion.

There is nothing that obliges the contractor to provide additional information requested by the architect but, obviously, if the architect reasonably asks for information to assist in making the decision the contractor would be foolish to withhold it. For example, the architect may say that it is not possible to understand the effects unless the contractor provides a critical path network showing the effect of the delay on the completion date. That does not appear to be an unreasonable request. On the contrary, it seems to be a request in the interests of all parties.

92   Deciding extensions of time is very difficult. Is there an easy way?

Most architects have difficulty in calculating extensions of time. In some instances the architect is reduced to trying to decide how much of the delay is the contractor's responsibility rather than how much of the delay is not its responsibility by virtue of the extension of time clause. Very often the decision is reduced to guesswork or, perhaps worse, how little can the architect get away with before the contractor seriously complains.

Obviously, deciding upon an extension of time is not a precise science. This is reflected in SBC which refers to 'ascertaining' (finding out for certain) loss and/or expense, but only to a 'fair and reasonable' extension of time. However, that is not to say that the architect can simply guess. That is certainly not acceptable.[135]

A useful tool is the computer planning program. Contractors find it helpful in preparing programmes for construction works and architects can use it to analyse the effects of delays. The computer program analyses the construction programme in the form of a network, critical path analysis or precedence diagram (sometimes referred to as a PERT – Performance Evaluation and Review Technique – chart). All these charts do is provide a way of connecting together the operations on site in a logical way, showing that some activities are dependent upon others and that some can start before others are completed. Resources can be introduced and a specific work calendar designated for each project. All architects and project managers should use computerised programs to monitor progress

---

135 *John Barker Construction Ltd v London Portman Hotels Ltd* (1996) 12 Const LJ 277

and assist in analysing claims. Contractors should submit detailed programmes on disk as well as on hard copy as a matter of course.

The courts have shown themselves ready to accept such analysis. In *Balfour Beatty Construction Ltd v The Mayor and Burgesses of the London Borough of Lambeth*,[136] there were references to the use of programs for estimating extensions of time. As part of its submission to the adjudicator, the contractor referred to the 'most widely recognised and used' delay analysis methods:

(I) Time Impact Analysis (or 'time slice' or 'snapshot' analysis). This method is used to map out the impacts of particular delays at the point in time at which they occur permitting the discrete effects of individual events to be determined.

(II) Window analysis. For this method the programme is divided into consecutive time 'windows' where the delay occurring in each window is analysed and attributed to the events occurring in that window.

(III) Collapsed as-built. This method is used so as to permit the effect of events to be 'subtracted' from the as-built programme to determine what would have occurred but for those events.

(IV) Impacted plan where the original programme is taken as the basis of the delay calculation, and delay faults are added into the programme to determine when the work should have finished as a result of those delays.

(V) Global assessment. This is not a proper or acceptable method to analyse delay.

There are numerous methods of using computer programming to arrive at extensions of time, but perhaps the most common are the methods briefly set out in (III) and (IV) in the quotation above. These techniques are sometimes known as the 'subtractive' or 'additive' methods. If comprehensive records are available, in the subtractive method a programme can be prepared to show the as-built situation and delays attributable to relevant events can be taken out to reveal what would have happened if the delays had not occurred. The second – additive – method, which is probably not quite so accurate, is the reverse. In this, the contractor's original programme is subjected to the addition of all the delays attributable to relevant events. The results in both cases are not precise. They prob-ably represent the greatest extension of time to which the contractor is entitled, because, in practice, the contractor would be expected to use best endeavours to reduce the effect of the delays, perhaps by re-programming. The logic links will have the greatest effect on the result and, therefore, they must be properly represented.

The need to take into account revisions to the programme was recognised in the *Balfour Beatty* case where the court said that, for the contractor to challenge an extension of time:

> … the foundation must be the original programme (if capable of justification and substantiation to show its validity and reli-ability as a contractual starting point) and its success will similarly depend on the soundness of its revisions on the occur-rence of every event, so as to be able to provide a satisfactory and convincing demonstration of cause and effect.

Of course, there will be occasions when it is unnecessary to prepare complex computer programs. In simple cases, where there is per-haps just one cause of delay, the critical path may be obvious. In most other cases, however, an effort should be made to analyse the delays in logical fashion. With the advent of adjudication, chal-lenges to extensions of time are becoming more frequent and the instances when an architect is being required to explain how an extension of time was calculated are increasing.

93   SBC: The project manager agreed the length of an extension of time with the contractor. Should the architect now certify it?

It is assumed that the architect is so named in the contract and that the project manager has been appointed by the employer. The only person empowered under the contract to extend the time allowed for the contractor to complete the Works is the architect.

Unless also appointed under the building contract as contract administrator, the project manager appointed by the employer has no powers under the contract. At best, the project manager is simply the employer's representative if the employer so notifies the various parties associated with the contract (clause 3.3). For the project manager to agree an extension of time with the contractor is a seri-ous infringement of the architect's powers and the architect should not contemplate certifying unless, of course, the architect also agrees the length of the extension of time. That is because the certificate is a formal expression of the architect's opinion. Obviously, it is not the architect's opinion if it has been agreed by the project manager.

In these circumstances, the architect should write to all parties setting out very clearly that under the terms of the contract only the architect may make an extension of time. If the project manager's view of the extension of time is supported by the employer, the architect should seriously question whether he or she can continue as contract administrator.

94 Can the client legally prevent the architect from giving an extension of time?

Under standard forms of building contract, it is the architect's duty to give an extension of time to the contractor if the relevant criteria have been met. The building contractor and the employer have entered into a contract whereby they both have agreed that when the contract criteria are satisfied, the architect will give the contractor an extension of time. Design and build contracts such as the JCT DB contract are excluded from this generalisation of course, because no architect is involved.

It is common for an architect to notify the employer before an extension is given, but it is not strictly necessary and it could be said to be misleading. The employer may think that the architect is seeking consent to the extension. If the architect does refer the extension of time to the employer, it is probably wise to make clear that the notification is a courtesy and for information only. Having said that, it is difficult to criticise an architect who takes the widest possible soundings before deciding on an extension of time. Not only the employer, but also the clerk of works and other consultants could be canvassed. Usually the architect will be seeking factual testimony so that the length of particular delays can be established with accuracy. The essential thing is that the architect must not only decide the extension of time, but also make that quite clear even though others may be consulted.

In *Argyropoulos & Pappa v Chain Compania Naviera SA,*[137] the (Plaintiff) architect, under the JCT Contract for Minor Building

Works 1980, gave extensions of time to the contractor and the employer objected – even going to the extent of considering the extension of time clause 'no longer valid'. The employer refused to accept the extension and a later extension given by the architect. The architect was informed that employer approval was required for any extension of time. At one stage the employer visited site and told the contractor that the architect had no power to give extensions of time. Eventually, the architect, on the advice of solicitors, withdrew its services. The extension of time point was just one part of the case, but in relation to that the judge said:

> ... the Defendants sought to interfere with the Plaintiffs' performance of their duties under [the extension of time clause] which they very properly resisted. Some of [the Defendants'] letters were also very offensive and indicated a total lack of confidence in the Plaintiffs. [The Defendants and their] Solicitors also undermined the Plaintiffs' position in relation to the contractors. In my judgment the Defendants' letters, the Solicitors' letters and the Defendants' conduct were in breach of contract and the Plaintiffs were amply justified in treating their engagement as at an end.

Not only does that show that interference with the architect's duty to give extensions of time is unlawful, in some circumstances, it probably amounts to repudiation on the part of the employer.

95 The employer terminated the contractor's employment in the 9th month of a 10-month SBC contract. The contractor is now claiming 16 weeks' extension of time.

Clause 2.28 provides that, if the contractor notifies the architect that the Works are being or are likely to be delayed and if this is done within a reasonable time of the delay or likely delay becoming apparent, the architect must give an extension of time. A key criterion in clause 2.28.1.2 is that the delay is likely to cause the date for completion in the contract to be exceeded. If the contractor's employment is terminated before the date for completion, there is no chance that it will be exceeded, therefore, no extension of time can be given. As a matter of plain common sense, the purpose of giving an extension of time is to give the contractor more time in which to complete the Works. Once the contractor's employment is terminated, there is no need for more time. However much time the contractor is given, it cannot complete before the date for completion, because it is no longer working on the site. Therefore, the contractor is not entitled to, and the architect has no power to give, any extension of time in these circumstances.

Why is the contractor seeking an additional 16 weeks' extension of time? One could see the point if the contractor was already in a period of culpable delay; it would reduce or remove liquidated damages. The contractor is probably looking for an extension of time in this instance because it misunderstands the provisions of the contract. Many contractors believe that an extension of time is necessary before a claim can be made for direct loss and/or expense. That is a completely wrong view. If the contractor believes it is entitled to loss and/or expense under clause 4.23 of the contract it must make an application under that clause. Loss and/or expense does not depend on extensions of time at all. If an extension of time is given to the contractor for a reason that is also ground for loss and/or expense under clause 4.4.24, the contractor can obviously use that extension as part of the evidence under clause 4.23, but a separate application must be made. But there is no reason why a contractor cannot make application for loss and/or expense without an extension of time. All that is required under clause 4.23 is that the contractor satisfies the criteria, none of which refer to extensions of time.

96   SBC: Is it permissible for the architect to give a further extension of time if documents from the contractor have not been received until after the end of the 12 weeks' review period?

Many commentators say that the architect is not bound by the 12 weeks' period because the period is not mandatory, but only directory on the authority of the Court of Appeal in *Temloc Ltd v Errill Properties Ltd.*[138] Commentators who take this view may perhaps not have read the judgment with the care it deserves. A careful reading shows that the court in *Temloc,* in saying that the 12 weeks was not a mandatory period, were actually interpreting the provisions

138 (1987) 39 BLR 30

against the employer that was seeking to rely upon them (the *contra proferentem* rule).

In that case, the employer had stated '£ nil' as the figure for liquidated damages and the Court of Appeal held that, if the contractor did not complete the work until after the date for completion, liquidated damages would be chargeable only at the stipulated rate, which would amount to no liquidated damages at all. The court said that the employer could not decide to claim unliquidated damages instead. In the usual way, the contract provided that after practical completion the architect must, within 12 weeks, confirm the existing date for completion or fix a new date. The architect did not act within the 12 weeks and the employer's position was that the liquidated damages clause could be triggered only if the architect carried out the duty within the 12 weeks. Therefore, the employer asserted the right to claim unliquidated damages for breach of an implied term. It was in this context that the court said that the time period was not mandatory.

The court apparently accepted the architect as the employer's agent. The matter could have been cleared up very simply on the basis that if the employer's argument succeeded, it would have been contrary to the established principle that a party to a contract cannot take advantage of its own breach.[139]

In a more recent case, the 12-week review period was confirmed in the following paragraph: 'The process of considering and granting extensions of time is to be completed not later than 12 weeks after the date of practical completion …'.[140] Therefore, it follows that if the contractor submits information after the 12 weeks has expired, the architect cannot consider that information. That is the case even if it is clear that if the information had been provided on time an extension of time would have resulted. To avoid this kind of difficulty, or at any rate to avoid any uncertainty, it is good practice for the architect to write to the contractor shortly after practical completion with a reminder about the deadline and requesting any further information that the contractor wishes to be considered no later than, say, week 7 of the 12. That puts the contractor on notice and, if the information is not provided on time, the contractor has no one else to blame.[141]

139 *Alghussein Establishment v Eton College* [1988] 1 WLR 587 HL
140 *Cantrell v Wright & Fuller Ltd* (2003) 91 Con LR 97
141 *London Borough of Merton v Stanley Hugh Leach Ltd* (1985) 32 BLR 51

Obviously, sometimes there are pressing reasons why it is desirable that the architect considers late submissions, for example when to do otherwise would be to risk time becoming at large. In these hopefully isolated cases, the parties to the contract can agree to give the architect power to consider the matter after the 12 weeks has expired. That is because the parties to a contract can always agree to vary the terms of their contract if they wish.

97 The contractor has a 4 weeks' extension of time. Can the employer charge for supplying electricity during this period?

The answer to this question depends on what it says in the specification or bills of quantities. If the employer is charging for the supply of electricity to the contractor during the contract period, then there is no good reason why the charge should not continue to be made during any period of extension of time.

If the employer is supplying electricity free during the contract period, a charge can be made if there is an end date specified and the extension goes beyond that date. If the employer has simply undertaken to supply electricity free or free during the contract period or free until the completion date, no charge can be made for the extension of time period.

Of course, if it is not being supplied free of charge by the employer, there is no reason why the contractor should not attempt to recover the cost of electricity during the 4 weeks if the grounds for extension of time are also grounds for claiming loss and/or expense and if the contractor makes an application under the relevant loss and/or expense clause in the contract. However, the employer should not pre-empt the situation by waiving the charge before the architect has made a decision on loss and/or expense (if claimed).

16 Liquidated damages

98 Is there a time limit for the issue of the certificate of non-completion under SBC and IC?

The certificate of non-completion is governed by clauses 2.31 and 2.22 in SBC and IC respectively. These clauses provide that the certificate must be issued by the architect if the contractor fails to complete the Works by the contract date for completion or any extension of that date. There is no express stipulation that the certificate must be issued by any particular date, although it is surprising how many people believe that it must be issued within 7 days of the contractor's failure to complete. This is incorrect. The only time limit is that imposed by the issue of the final certificate. The final certificate is the architect's final action under the contract. After issuing it the architect is *functus officio* – that is to say the architect has no further powers or duties and, therefore, cannot issue the non-completion certificate.

99 The employer terminated in the 9th month of a 10-month contract. Can the employer deduct liquidated damages from the original contractor until practical completion is achieved by others?

In general terms it appears that, if the employment of the contractor is terminated, the obligations of both parties is at an end in so far as future performance is concerned.[142] This seems to be perfectly in accordance with good sense, because if the Works are completed by another contractor the original contractor can have

---

142 *Suisse Atlantique etc v N V Rotterdamsche Kolen Centrale* [1966] 2 All ER 61

no control over the completion. That is not to say that a party will avoid the payment of damages accrued up to the time of termination.[143]

The decision in *re Yeardon Waterworks Co & Wright*[144] suggests that the courts will support a specific term in the contract that provides that in the event of termination of the employment of a contractor and the completion by another, damages could be deducted until the Works are completed. In that case, however, the Works were completed by the guarantor of the contractor, which was probably the deciding factor.

The JCT series of contracts provide for termination of the contractor's employment, following which the employer may engage another contractor to enter site and complete the Works. Such a clause was held to be incompatible with the right to liquidated damages in *British Glanzstoff Manufacturing Co Ltd v General Accident Fire & Life Assurance Corporation Ltd.*[145] If a contractor has left the site, wrongly thinking that the Works are complete, it seems that contractor will be liable for liquidated damages until the Works have in fact been completed by a replacement contractor.[146] The precise wording of the clause in the contract will be the deciding factor. In the New Zealand case of *Baylis v Mayor of the City of Wellington*,[147] liquidated damages were held to be deductible after termination, because the clause specifically excluded entitlement during the time taken by the employer to secure a replacement contractor.

In *re White*,[148] the electric lighting contract contained what was held to be a liquidated damages clause. The court remarked that there was a clause in the contract which gave the engineer power, if necessary, to employ other contractors to complete the Works, and provided that the defaulting contractor should be liable for the loss so incurred without prejudice to his obligation to pay the liquidated damages under the contract. It is not clear from the report whether the employer was seeking liquidated damages beyond the date of termination. The employer does not, however, appear to have claimed anything other than liquidated

---

143 *Ex parte Sir W Harte Dyke. In re Morrish* (1882) 22 ChD 410
144 (1895) 72 LT 832
145 [1913] AC 143
146 *Williamson v Murdoch* [1912] WAR 54
147 (1886) 4 NZLR 84
148 (1901) 17 TLR 461

damages, despite the words of the contract, which appear to give the employer the right to claim liquidated damages for breach of obligation to complete on time until the date of actual completion, together with all the additional costs associated with completion by another contractor.

The effect of termination on the right to recover damages was considered in *Photo Production Ltd v Securicor Transport Ltd.*[149] Speaking of another case,[150] it was said:

> ... that when in the context of a breach of contract one speaks of 'termination' what is meant is no more than that the innocent party or, in some cases, both parties are excused from further performance. Damages, in such cases, are then claimed under the contract, so that what reason in principle can there be for disregarding what the contract itself says about damages, whether it 'liquidates' them or limits them, or excludes them?

This seems to be a clear reinforcement of the view that there can be no continuing liability to pay liquidated damages, but damages already accrued are recoverable. Standard forms of building contract state the grounds on which either party may terminate the contractor's employment under the contract. Many of the grounds for termination under the provisions of the contract are not breaches that would entitle the employer to terminate save for the express provision. It thought that an employer who terminated using the contract provisions is restricted to recovering the amounts stipulated in the contract.[151] Current building contracts do not appear to allow the continued deduction of liquidated damages after termination. In any event, the circumstances set out in the question suggest that, even if the contractor's employment was not terminated, liquidated damages would not be due until a further month had passed, because at the date of termination the date for completion had not been reached.

149 [1980] 1 All ER 556
150 *Harbutt's Plasticine Ltd v Wayne Tank and Pump Co Ltd* [1970] 1 All ER 225
151 *Thomas Feather & Co (Bradford) Ltd v Keighley Corporation* (1953) 52 LGR 30

100 Can actual damages be claimed instead of liquidated damages if overrun is very long?

The frustration of an employer can be appreciated when what seemed to be a simple, short building contract runs on and on as though it will last for ever. There are many instances of 3-month contracts for the addition of a small extension to a domestic property extending to more than twice their original contract periods.

Liquidated damages are not a punishment for a contractor who is late. If they were inserted with that in mind, they would become a penalty and unenforceable against the contractor. Rather they are inserted to recompense the employer for the anticipated costs of overrun as calculated at the time the contract was entered into.

Often, the amount inserted as liquidated damages is too small to properly reimburse the employer for the loss of being kept out of the building. The argument is that to include an adequate sum in the contract would dissuade from tendering just the kind of small-ish builders who tend to specialise in this sort of work. But the problem is that the liquidated damages is the whole of what can be claimed by the employer as a result of the contract completion date being exceeded.[152] Therefore, even if the 3-month contract overruns by a year or more, only the amount of liquidated damages in the contract can be recovered per week of overrun.

Of course, there is absolutely nothing to prevent the employer from structuring the amount of liquidated damages payable so that, say, £y is payable every week for the first 4 weeks, then £y × 2 for a further 4 weeks, then £y × 3 and so on. Liquidated damages with much more complex structures than that have been allowed by the courts.[153] However, the employer must be able to justify the increase in liquidated damages if called upon to do so by the court or an adjudicator or arbitrator. For example, there must be evidence that after a certain date, the cost of storing furniture or rental costs will rise. In providing such justification, it is what the employer envisaged at the time of entering into the contract that is important, not any actual movement of costs later.

---

152 *Temloc Ltd v Errill Properties Ltd* (1987) 39 BLR 30
153 *Philips Hong Kong v Attorney General of Hong Kong* (1993) 61 BLR 41;
     North Sea Ventilation Ltd v Consafe Engineering (UK) Ltd, 20 July 2004 unreported

101 What does a contractor mean who says that 'damages are at large'?

This is a little-used, archaic kind of term. It is probably used mostly by contractors who are confused between time at large and liquidated damages.

To say that damages are 'at large' simply means that damages are to be assessed by the courts or an arbitrator as the case may be. In short, it is the usual situation in law. The significance of the phrase when used by a contractor is in contrast to liquidated damages.

The whole point of liquidated damages is that such damages have been agreed by the parties as a genuine pre-estimate of the loss likely to be suffered by the employer in the event of some pre-determined occurrence. In construction contracts, this is usually when the time for completion of the Works is exceeded. Therefore, liquidated damages refers to an agreed amount. Such damages are sometimes referred to as 'agreed damages'. Where no such damages are agreed, the amount that can be recovered is referred to as 'unliquidated damages' or one could say that damages are 'at large'.

In most standard form contracts, if the contractor is in breach of contract by completing the Works later than the contractual completion date or any extended date, the amount the employer can recover as a consequence of the breach is pre-determined as liquidated damages, usually a sum per day or per week. However, if the contractor is in breach of some other obligation under the contract, say for example, the contractor walks off site and refuses to come back, the employer will be entitled to bring an action at common law to recover whatever damages can be proved to have resulted from the breach. Those are damages at large, because they have to be proved, unlike liquidated damages which are already agreed.

When a contractor argues that damages are at large, it is because the contractor believes that will gain it some advantage. If liquidated damages are a considerable sum per week and the contractor in delay has little or no ground for an extension of time, the contractor may try to argue that some action of the employer has rendered the date for completion inapplicable. It may say that time is, therefore, at large and that liquidated damages can no longer apply, therefore damages are at large also. The contractor will argue in this way only if it seems clear that actual, unliquidated, damages will amount to less than the liquidated damages in the contract or if proving damages will be a difficult task for the employer. It is rare for a contractor to successfully argue that time is at large and consequently equally rare for the argument that damages have become at large to succeed.

102 If the employer has to cancel a £13,000 holiday because a new house is not completed in week 17 of a 14-week contract, can the employer claim the holiday cost? Liquidated damages are £250/week.

The purpose of inserting liquidated damages in the contract is to avoid the employer having to claim damages for the contractor's breach of its obligation to complete by the contractual date for completion. If it were necessary for the employer to claim damages through the courts when such a breach occurred, the employer would be faced with an expensive case, because the breach would have to be established and evidence brought to show the amount of damage the employer had suffered as a direct result of the breach.

Therefore, building contracts almost invariably adopt the system of liquidated damages to enable the employer to insert a sum in the contract which is to be payable by the contractor for every day or week by which the completion date is exceeded.

It is not clear from the question exactly why the employer has to cancel the holiday simply because the contractor is late. But even if the cancellation is a direct result of the contractor's delay, the employer cannot claim the holiday cost, because the amount inserted as liquidated damages is said to include all the financial effects of the delay.[154] If the employer knew about the holiday before the building contract was entered into, the liquidated damages should have been adjusted accordingly to take account of the possibility of a missed holiday. There is no difficulty with grading liquidated damages or in providing for much greater damages during a particular period. The only thing to remember is that, if such damages are challenged in adjudication or arbitration, it will be necessary to show that, at the time they were put in the contract, they were a genuine pre-estimate of the likely future losses.

In this instance, the employer is restricted to a maximum of £250 per week.

---

154 *Temloc Ltd v Errill Properties Ltd* (1987) 39 BLR 30

103 Can the employer still claim liquidated damages if possession of the Works has been taken?

Architects and contractors alike often labour under the mistaken impression that once the employer occupies the Works no further liquidated damages can be levied. Indeed, possession is often – wrongly – equated with practical completion. An examination of the definitions of practical completion given by the courts shows that the criteria are whether the Works, except for minor items, are complete and whether there are any visible defects. Whether the employer is or is not in occupation is not a factor.

The recent case of *Impresa Castelli SpA v Cola Holdings Ltd* [155] gives some useful guidance. The contract was the JCT Standard Form of Contract With Contractor's Design 1998. Disputes arose and the parties made no fewer than three separate variations to the contract terms. In particular, the liquidated damages amount was doubled and the date for completion was amended three times. Significantly, it was agreed that the employer (Cola) could have access to the hotel in order for it to be fully operational.

The contract finished late and Cola claimed liquidated damages of £1.2 million. Impresa challenged that amount, arguing that Cola had taken partial possession and, therefore, the liquidated damages should be considerably reduced. Significantly, the court decided that there was nothing to suggest that partial possession had occurred. It would have been quite simple to have referred to 'partial possession' in any of the three agreements if that is what had been intended, but there was no such reference. Instead, the court concluded that clause 23.3.2 (which was virtually the same as clause 2.5 of the current DB and clause 2.6 of SBC) had been operated, which allowed the employer to use and occupy the Works with the contractor's consent. The court decided that is a lesser form of physical presence on the site than possession. Therefore, the full amount of liquidated damages was recoverable by Cola.

Sometimes there is no agreement at all, but the employer, perhaps frustrated at continuing delays, simply decides to move in. The courts have held that such occupation did not preclude the deduction of liquidated damages.[156]

---

155 (2002) 87 Con LR 123

156 *BFI Group of Companies Ltd v DCB Integration Systems Ltd* (1987) CILL 348

104 SBC: If practical completion is certified with a list of defects attached, can the employer deduct liquidated damages until termination (which occurred later due to the contractor's insolvency)?

Clause 2.32.2 of the contract specifies that the employer may give notice that payment of liquidated damages is required or that liquidated damages will be deducted from the date on which the contractor should have completed the Works (or any section) until the date of practical completion. Practical completion marks the date on which the contractor has completed its obligation to construct the Works and, therefore, the employer is no longer suffering any damage due to non-completion after that date.

Although, in *Tozer Kemsley & Milburn (Holdings) Ltd v J Jarvis & Sons Ltd & Others*,[157] the judge, without criticism, refers to a schedule of defects added to the certificate of practical completion, it is established that a certificate of practical completion cannot be issued if there are known defects in the Works.[158]

Therefore, it is clear that the certificate of practical completion should not have been issued while there were known defects in the Works. In this instance, it would probably be open to the employer to seek adjudication or arbitration on the basis that the practical completion certificate had been wrongly issued. If the adjudicator or arbitrator agreed, the employer would be entitled to recover liquidated damages until the date of practical completion as properly certified. If termination due to the contractor's subsequent insolvency occurred after practical completion, it would be irrelevant so far as the calculation of liquidated damages was concerned. If the employer does not take this step, the certificate is the cut-off point for liquidated damages and, so far as the list of defects is concerned, it would be subsumed into defects occurring during the rectification period.[159]

157 (1983) 4 Con LR 24
158 *Westminster Corporation v J Jarvis & Sons* (1970) BLR 64
159 *William Tomkinson and Sons Ltd v The Parochial Church Council of St Michael* (1990) 6 Const LJ 319

105 If the employer tells the contractor that liquidated damages will not be deducted, can that decision be reversed?

More often than one might expect, an employer will tell a contractor that liquidated damages will not be deducted either totally or for part of the contractor's delay. This is usually done because a contractor is complaining bitterly about life in general and the contract in particular and the employer is worried that, if faced with the prospect of huge liquidated damages, the contractor may simply walk away from the project. The contractor would be wrong to do that and liable for damages, but completing the project with a fresh contractor and trying to recover damages from the old contractor would be time-consuming and traumatic.

Sometimes an employer has been known to state that liquidated damages will not be deducted instead of the architect giving an extension of time that is clearly due. That kind of conduct is asking for trouble. It derives from the false notion that, if the contractor is not given an extension of time, it cannot subsequently make a claim for loss and/or expense. The actual result is that time probably becomes at large and the contractor can still make an application for loss and/or expense if it is so minded. One thing is certain: the employer should not say that liquidated damages will not be deducted and then undergo a change of mind.

That is what happened in *London Borough of Lewisham v Shephard Hill Civil Engineering*.[160] In that case, the contract was the ICE 6th edition, but the principle holds good for JCT contracts also. The case was actually an appeal to the court from the decision of an arbitrator. It was alleged, before the arbitrator, that Lewisham or its engineer had assured Shephard on several occasions that liquidated damages would not be recovered. Apparently these assurances were given orally and not confirmed in writing. At the time of the arbitration the liquidated damages stood at a considerable sum – about £550,000. Shephard contended that, as a result of these assurances, they had paid their sub-contractors without deducting damages which otherwise they would have done. The arbitrator noted that the engineer's final extension of time was not given until about 2 years after substantial completion of the Works and Lewisham did not claim the

---

160   30 July 2001 unreported

liquidated damages until after the arbitration had begun and was by way of a counterclaim.

In all the circumstances, the arbitrator concluded that Lewisham did make the representations about liquidated damages. The arbitrator, therefore, held that Lewisham was estopped (prevented) from claiming liquidated damages, the basis of that decision being the well-known principle that if one party to a contract makes a representation to the other that it will not enforce particular terms of the contract and if the other party relies on that representation to its detriment, the first party will be estopped from later trying to enforce that particular term. Shephard had relied on Lewisham's representation to its detriment when it had paid its sub-contractors without deduction. The court held that the arbitration had jurisdiction to decide the issue.

On the assumption that the representations were given as alleged, the decision was clearly correct. An important point was that Shephard had relied on the representation when deciding to pay its sub-contractors without deduction. That was a very clear-cut case of reliance. Even if payment of sub-contractors was not an issue, most contractors would be able to show that they had acted differently as a result of a representation not to deduct liquidated damages.

106   The contract shows the completion date of a school, but the contractor has submitted a programme showing a later completion date. From which date do liquidated damages run?

The first thing to get clear is that few forms of contract, and none of the JCT series, make the programme a contract document. Submission of a programme is a requirement under clause 2.9.1.2 of SBC, but it does not become a contract document. If it did, all parties would be required to work from it, perhaps with unexpected results.

Under most forms of contract, the contractor will submit a programme, whether requested or not, and it is sensible to do so. However, even if the programme is approved by the architect, project manager or the employer, the programme remains the creation of the contractor and something for which the contractor retains responsibility.

The contractor's obligation to complete is set out in the contract. For example, in SBC, clause 2.4 requires the contractor to regularly and diligently proceed with and complete the Works or section on or before the relevant completion date. Liquidated damages are dealt with in clause 2.32 and depend for recovery on the issue of a non-completion certificate by the architect. Clause 2.31 deals with this certificate, which is to be issued by the architect if the contractor fails to complete the Works or a section '*by the relevant Completion Date*'. The date for completion of the Works or sections is to be inserted in the contract particulars.

Therefore, it is clear that it is the date in the contract which is important. This date can be changed only in two ways. The first, and fairly unusual, way is if the parties to the contract agree to vary the contract terms by amending the date. The second, and more usual, method is if the architect gives an extension of time under the provisions of the contract. The contractor's programme cannot change the completion date; that is so even if the programme with the later date is approved by the architect. In such circumstances, 'approval' would be a strange thing for the architect to give, but only the employer and the contractor, not the architect and the contractor, have the power to vary the contract in this way. The architect is agent for the employer in a very limited way, which certainly does not include acting for the employer to vary the contract terms. A contractor who has tendered for the Works on the basis of contract, drawings and bills of quantities or specifications and a period for carrying out the Works and who has executed a formal contract to that effect should not be allowed to submit a programme showing a later completion date. The architect should make clear that the contract must show how the contractor intends to complete on or before the completion date.

From the foregoing, there should be no doubt that liquidated damages would not run from the programmed date for completion, but from the contract date or any proper extension of that date under the terms of the contract.

17 Loss and/or expense

107 It is 3 months after practical completion and the contractor has just produced a claim in four lever arch files. What should be done about it?

Under clause 4.23 of SBC and clause 4.17 of IC, the contractor's entitlement to loss and/or expense is made subject to a written application having been made to the architect within a reasonable time of it becoming apparent that the contractor is incurring or is likely to incur direct loss and/or expense for the grounds laid down. SBC actually goes somewhat further and makes the requirement a proviso or condition.

What is a reasonable time? What is reasonable is always a difficult question to answer, because the standard (and correct) answer is that it depends on all the circumstances. So far as a claim submitted 3 months after practical completion is concerned, it appears at first sight as though that is not within a reasonable time.

The first thing to be done is to examine the claim to identify the grounds and to relate the grounds to the times when the contractor knew that direct loss and/or expense was being suffered. That should be relatively easy to do. For each item, the architect must work out the date when the contractor must have known that it was either incurring or likely to incur the loss and/or expense. The next point to establish for each item is whether the contractor did make an application to the architect. In this context, it is suggested that the making of an application can be given quite a broad interpretation. Therefore, it might be enough if the contractor merely refers to a future claim for loss and/or expense for the item in question. The contractor is possibly giving notice of delay for the purpose of extension of time and may say something like 'and we shall require associated loss and expense'. The fact that loss and/or expense is not associated with extension of time should be ignored for this purpose. Or the contractor might say that it has received the architect's instruction and that it is likely to cause loss and/or expense. The point is whether the architect was given warning or, to use the legal phraseology, 'put on notice' that a claim for loss and/or expense was likely.

The whole point of the requirement for the contractor to make application within a reasonable time is to give the architect the opportunity to take steps at a time when there was still the opportunity to do so. For example, the architect may well have decided to cancel an architect's instruction (if that was the cause of the problem). It may be that other measures could have been taken by the architect or the employer to reduce the amount of likely loss and/or expense. Perhaps most importantly, the architect could have taken steps to ensure that adequate records were kept of the contractor's work and times. It is important to understand this in order to decide whether the contractor acted promptly. If a contractor knew immediately it received an architect's instruction that loss and/or expense would be incurred, the application should have been made immediately. If the contractor realised the problem only when it was in the process of carrying out the relevant work, that is the time when it would have been reasonable to make the application. The application must identify the reason for the application, including the circumstances and the contract clause on which the contractor relies. Further information can be sent later. The important thing is that sufficient information is given to the architect to enable the basic claim to be identified. There is no obligation for the contractor to calculate the amount of the claim at this or any other time.

The phrase 'within a reasonable time' does not give the contractor freedom to wait for several months until notifying the architect. A reasonable time is clearly as quickly as practicable having regard to the contractor's own knowledge. If the contractor is late in making application and appears to be outside the time frame envisaged by 'a reasonable time', it is suggested that the final test to be applied is whether the employer will be prejudiced by the late application. For example, an employer could be prejudiced if the lateness of the application made it difficult for the architect to ascertain the facts or in some other way was unable to operate the machinery of the claims clause correctly.

In the case of the claim submitted 3 months after practical completion, any items that have already been the subject of a timely application, even if brief, must be identified and be considered. So far as the others are concerned, the application was made at least 3 months after the disruptive or delaying events were known to the contractor. It is difficult to escape the conclusion that the architect has no power under the contract to consider them. That does not mean that the contractor has lost any remedy for these items. If they are also breaches of contract or other defaults, taking action at common law for damages is still available.[161] Of course, it is always open to the employer to agree with the contractor that, even if late, the architect will consider the application. In this instance, the architect must make sure to notify the employer in writing of the ways in which the employer may be prejudiced by the late application and obtain the employer's express instruction to proceed.

Once this initial point has been established, the architect can proceed to consider the claim in the usual way: by checking that the contractor has identified the contract clauses and relevant matters on which it relies; by establishing that the facts support the contention that there has been an occurrence that falls under the relevant matter; that the occurrence has resulted in direct loss and/or expense for the contractor; and that the contractor would not be reimbursed by any other payment under the contract. It should be noted that the contract does not refer to the contractor having been reimbursed, but only that it would have been reimbursed. Therefore, the point is not whether the contractor has been paid, but whether the contractor was entitled to be paid.

108 The contractor is demanding to be paid 'prelims' on the extension of time. How is that calculated?

None of the standard forms of contract entitle the contractor to any payment as a consequence of extension of time. The purpose of extension of time clauses is that the period of time available for carrying out the contract Works can be extended. The contractor will look in vain for any reference in the contract to money in connection with extension of time. In short, there is no connection

---

161 *London Borough of Merton v Stanley Hugh Leach Ltd* (1985) 32 BLR 51

between the contractor's entitlement to an extension of time and any entitlement to loss and/or expense.[162]

There is no provision for the payment of 'prelims' in any of the standard forms. 'Prelims' is short for 'preliminaries' i.e. the first part of the bills of quantities or specification. When a contractor talks about recovering its preliminaries it means the price for the preliminary items that has been inserted in the bills or specification as part of the tender price. This preliminaries price is often inserted as a lump sum or as a price per week. Obviously a lump sum can easily be converted into a weekly rate, by dividing it by the number of contract weeks.

It was once very common, but now fortunately less so, for the architect to give a contractor an extension of time and then for the quantity surveyor to ascertain the amount payable to the contractor by multiplying the weekly amount of preliminaries in the specification by the number of weeks' extension of time. As construction professionals better understand their responsibilities in dealing with loss and/or expense, this practice is less common. It borders on negligence unless sanctioned by the employer in the full knowledge of all its implications. The contractor can never be entitled to recover its preliminaries as loss and/or expense, much less as a rate per week of extension of time, because the contractor is entitled to recover only its actual losses or actual expenses – in other words, the amount it can prove it has actually lost or spent.[163] It is not entitled to recover some notional amount, nor the amount inserted as part of its tender price which may, but more likely may not, be the same as its actual costs.

Therefore, the answer to this question is:

1. The contractor is not entitled to any money as a consequence of being given an extension of time.
2. Therefore, there is no need to calculate it.
3. If the contractor makes a valid application under the claims clause (e.g. clause 4.23 of SBC or clauses 4.17 and 4.18 of IC), it is entitled to the actual amount of loss and/or expense it has suffered.

---

162 *H Fairweather & Co Ltd v London Borough of Wandsworth* (1987) 39 BLR 106; *Methodist Homes Association Ltd v Messrs Scott & McIntosh*, 2 May 1997 unreported

163 *F G Minter Ltd v Welsh Health Technical Services Organisation* (1979) 11 BLR 1; (1980) 13 BLR 7 CA

109 Is it true that a contractor cannot make a loss and/or expense claim under MW?

Strictly speaking, that is correct. The only clause in MW that mentions loss and/or expense is clause 3.6.3. It provides that if the architect issues an instruction requiring an addition to or omission from or any other change in the Works or the order or the period in which they are to be carried out and there is a failure to agree a price before the contractor carries out the instruction, it must be valued by the architect.

The architect is required to value it on a fair and reasonable basis using any prices in the priced document. Significantly, the valuation must include any direct loss and/or expense that the contractor has incurred as a result of regular progress of the Works being affected in either of two ways. The first is compliance by the contractor with the instruction. The second is the employer complying or failing to comply with clause 3.9.

Clause 3.9 requires both parties to comply with the CDM Regulations. In particular, under clause 3.9.1, the employer must ensure that the duties of the planning supervisor are properly carried out and, if the contractor is not acting as the principal contractor, that the duties of the principal contractor are also properly carried out. The grounds on which the architect can include loss and/or expense are obviously quite restricted.

It is clear from the wording of clause 3.6 that the architect's inclusion of loss and/or expense in the valuation does not depend on any application by the contractor. Indeed, the only time the contractor is expressly required to provide information to the architect is under clause 4.8.1, where the contractor must provide all documentation reasonably required for computation of the final certificate.

In calculating the valuation, the architect will no doubt ask the contractor for information. Indeed, in practice most contractors will provide information in the form of an application for payment on a monthly basis. Although the contract does not preclude such applications, it does not confer any status upon them. The architect may take notice of or ignore the information as the architect deems appropriate, because the only factor the architect needs to take into account is the priced document, whether that is a priced specification, work schedules or a schedule of rates. It is entirely a matter for the architect how the loss and/or expense is calculated. Many architects link the amount to the length of any extension of time that has been given on account of architect's instructions. Although one can see some logic in this approach, there is no justification for arriving at the loss and/or expense by multiplying the number of weeks by the amount the contractor has inserted in the priced document as its weekly preliminaries cost. Loss and/or expense is the equivalent of damages at common law. As such, the damages must be proved; the architect must secure the necessary evidence to show how much loss the contractor has actually incurred. At best, the preliminaries figure in the priced document is the contractor's best estimate at tender stage. It may be an under- or an overestimate. It certainly will not represent actual costs.

The contractor cannot make a claim for loss and/or expense under the terms of MW, because they contain no mechanism to enable it to do so. MW does not have the equivalent of clauses 4.23 and 4.17 of SBC or IC respectively. Therefore, the contractor cannot make a claim under the contract for loss and/or expense for information received late. All is not lost, however. There is absolutely nothing to stop the contractor making a claim at common law in such cases, basing the claim on a breach of contract by the employer and claiming damages. Such action has received judicial blessing.[164] The architect cannot deal with such claims, because they are outside the contract machinery. They must be handled by the employer. If the architect receives such a claim, it must be forwarded to the employer immediately. The architect should refrain from expressing any view about the claim unless consulted by the employer. Theoretically, the employer should deal with the matter by separate legal advice.

---

164 *London Borough of Merton v Stanley Hugh Leach Ltd* (1985) 32 BLR 51

18 Sectional completion

110 Can sectional completion be achieved by inserting the phasing dates in the bills of quantities Preliminaries section?

It used to be a feature of local authority housing contracts that the bills of quantities contained a series of dates or times when certain blocks of dwellings had to be completed. Sometimes this was expressed as a number of weeks from the date of possession. For example: 'Block A – six weeks, Block B – ten weeks' etc. A variant was where existing properties were involved and the contractor was to be allowed possession of blocks of ten dwellings at a time and not entitled to take possession of another dwelling until one had been completed. If JCT contracts were involved, this process was unlawful. In the case of other types of contract, it might be lawful, but probably unworkable in practice.

Where there is just one date for possession and one date for completion in the contract, sectional possession or completion cannot be achieved by simply inserting intermediate dates in the specification or bills of quantities.[165] Certainly not in JCT contracts, which have a clause giving priority to the printed form over other contract documents (see, for example, clause 1.3 in SBC and IC). In these contracts, the single completion date in the contract particulars of the printed form will take precedence even though there is a whole regimen of blocks and dates in the bills of quantities, because the bills of quantities are not allowed to override or modify the printed form. Therefore, although the bills clearly state

---

165 *M J Gleeson (Contractors) Ltd v Hillingdon Borough Council* (1970) 215 EG 165

that the contractor can only take possession of houses in ten-dwelling lots, on the true construction of the contract the contractor's right is to take possession of the whole of the dwellings at the date for possession in the contract particulars and its obligation to complete does not bite until the single date for completion. If the dates in the bills of quantities have any relevance at all, which is doubtful, they may simply indicate the order in which the blocks are to be finished. However, a contractor completing all the dwellings on the single completion date has fulfilled its obligations.

If the contract is not part of the JCT series and there is no similar priority clause, the usual rule that 'type prevails over print' will apply and any sectional possession or completion dates stated in the bills of quantities or specification would apply in preference to the date in the contract. However, this would cause some contractual problems. For example, most standard form contracts provide for only one certificate of practical completion, one certificate of making good defects and one rectification period (or their equivalents). Such contracts refer to extension of time in relation to the single completion date or fixing a new single completion date. Such problems can be resolved only by goodwill on both sides or with the assistance of an adjudicator, arbitrator or judge. In the JCT 98 series of contracts, putting sectional completion into a standard building contract required a considerable number of contract amendments. Although this is no longer necessary in the 2005 series, because completion in sections has been incorporated into the basic contracts, where JCT contracts are not used a tremendous amount of amendment would be necessary to avoid problems – as a glance at the former sectional completion supplements will demonstrate.

111 The contract is SBC, which includes provision for sections. The employer wants to rearrange the sections. Can that be done with an architect's instruction?

The fact that the contract is to be carried out in sections has been agreed between the employer and the contractor when they executed the contract. In other words, the sections are a term of the contract. Therefore, in order to change the sections it is necessary to have a further agreement between employer and contractor. It is not something the employer can unilaterally decide, any more than the employer can unilaterally decide to reduce the contract sum by 20 per cent.

Still less can the sections be changed by the architect through the medium of an architect's instruction. Apart from any other consideration, the architect is not empowered by the contract to issue an instruction to that effect. Therefore, any such instruction would be void.

If the employer wishes to rearrange the sections, the contractor's consent must be sought. Even where both parties agree, the change cannot be achieved by an architect's instruction. If the contractor is willing, the employer must organise the drafting of a special addendum to the contract setting out the variation agreed between the parties and any other matters that arise from the change (for example, it will be necessary to amend the liquidated damages). Both parties must sign the addendum.

112 The contract is SBC in sections. The dates for possession and completion have been inserted for each section. Section 2 cannot start until section 1 is finished. The contractor is saying that possession of section 2 must be given on the due date even if it is the contractor's own fault that section 1 is not finished.

This is a common problem when the contract is divided into sections, each with its own date for possession and completion, but two or more of the sections are interdependent. For example, a refurbishment project may be divided into three sections, but section 2 may be dependent on section 1 in a practical sense, because the contractor cannot physically be given possession of section 2 until section 1 has been completed. That is usually because the occupants of section 2 have to be moved to section 1 when it is finished. Usually, the dates for possession and completion of each section are inserted into the contract as a series of dates. The date for completion of section 1 and the date for possession of section 2 will probably be separated by a week or so to allow occupants and furniture to be moved from one section to another.

If section 1 is not finished by the completion date, even due to the contractor's fault, the contractor is still entitled to take possession of section 2 on the appointed date in the contract particulars (see clause 2.4 of SBC). If it is physically impossible for such possession to take place, the employer will be in breach of contract. The contractor is, therefore, correct.

Where a project is split into sections, any extensions of time must be given in respect of the particular section affected by the delaying event. There is no provision that the delay in one section will affect another. Therefore, even if the whole of the delay to section 1 entitles the contractor to an extension of time, it will be only section 1 that is extended and not section 2.

If the cause of the delay is entirely the fault of the contractor, the architect may say that the contractor, being responsible for the delay to section 1, is clearly responsible for the delay to section 2 also and it cannot expect to take possession of section 2 on the date set out in the contract particulars. This approach is very common, but wrong. The cause of the delay to possession of section 2 is not the contractor's delay to section 1, but the fact that the two sections are linked. If they were not linked, the contractor's delay to section 1 would not affect section 2. One of the difficulties is that, where the dates for possession and completion are simply set out as a series of dates, there is nothing to warn the contractor about the likely problem.

A situation that was similar, but quite different in effect, occurred in *Trollope & Colls Ltd v North-West Metropolitan Regional Hospital Board*.[166] There were three sections, each with a separate contract sum and set of conditions. Although the start of section 3 was shown in the contract as being subject to the completion of section 1, the date for completion of section 3 was given as a particular date. Therefore, when the completion of section 1 was delayed, the start of section 3 was also delayed, but the date for completion of section 3 remained the same. Therefore, the period for completion was reduced from 30 to 16 months. This seriously affected the contractor, but the House of Lords refused to imply a term into the contract that the completion date for section 3 should be extended accordingly.

There are two probable supplementary issues: the first is whether anything can be done to avoid the problem in new contracts; the second is whether anything can be done where the situation outlined in the question is currently in place.

To avoid the problem is relatively straightforward. The employer must clearly show the links in the sections. The contract particulars

166 (1973) 9 BLR 60

should be amended to delete the current setting out against 'Sections: Dates of Possession of sections' and in its place or on a separate, but properly attached and signed, sheet, section 1 would have a date for possession and a date for completion, but section 2 would not have a date for possession. It would simply state: 'The date for possession is x days after the date of practical completion of section 1.' Therefore, a delay to completion in section 1 (from whatever cause) would be reflected in the date of possession of section 2 and there would be no breach of contract, because section 2 could be given to the contractor on the due date. The date for completion of section 2 would not be inserted, but rather: 'The date for completion is x weeks after the date that possession of this section was taken by the contractor'. It is difficult to see the grounds on which the contractor could make any financial claim on the employer for delays to section 1 that cause a delay to the possession of section 2 if this method of setting out the dates was implemented. It has been said that the dates for possession and completion cannot be entered in this way, because they are not actual dates. That would be to take the wording too literally. The important thing is that the wording enables the dates to be unerringly calculated, albeit not until practical completion of section 1 has taken place.

How to rectify the situation if proper provision has not been made is slightly more complex. If there is provision for the employer to defer possession of any of the sections by the appropriate amount, the employer must do so and the contractor will be entitled to an extension of time and probably whatever amount of loss it has suffered as a result of the deferment of possession. If there is no deferment provision or if the delay exceeds the period of deferment allowed under the contract, the situation appears to be that there is a breach of contract which, dependent upon circumstances, may be a repudiation. The contractor would be entitled to recover as damages the amount of loss he has suffered. If there was no provision for the delay situation in the contract, the architect would be unable to make any extension of time and the contractor's obligation with regard to section 2 would be to complete within a reasonable time. Therefore, liquidated damages for this section would not be recoverable.

SBC and IC now provide for such breaches to be dealt with by extension of time and loss and/or expense under SBC clauses 2.29.6 and 4.24.5 and IC clauses 2.20.6 and 4.18.5 respectively. The amount payable to the contractor, whether by virtue of a loss and/or expense clause in the contract or as damages for the breach may not be substantial. The contractor would have to demonstrate a loss and the situation is simply that section 2 has been pushed back in time.

19 Practical completion and partial possession

113 If the architect has issued a certificate of practical completion with 150 defective items listed and the contractor is not remedying them within a reasonable time, what can be done about it?

The leading case on the requirements for practical completion[167] states that practical completion cannot be certified if there are known defects in the Works. Therefore, a certificate issued with 150 defective items listed is, on its face, void or, perhaps more accurately, voidable if either party applies to an adjudicator. An adjudicator ought to find that the certificate was not properly issued and the contractor would be obliged to rectify the defects before a certificate could be properly issued.

If neither party seeks adjudication on the matter, and in practice the contractor is unlikely to do so because the certificate carries various advantages, the defects will become part of the defects to be dealt with during the rectification period.

If the contractor does not rectify within a reasonable time, the architect may issue an instruction requiring the defects to be made good followed by a compliance notice under SBC clause 3.11 or IC clause 3.9. If the contractor does not comply within 7 days, the employer may engage others and all the additional costs incurred by the employer will be deducted from the contract sum. Alternatively, if the employer is content to wait until after the end of the rectification period, the defects can be added to the schedule of defects and dealt with in the usual way.

---

167 *Westminster Corporation v J Jarvis & Sons* (1970) BLR 64

114 Is the contractor entitled to a certificate of practical completion after termination?

The certificate of practical completion indicates that the contract Works are almost complete, that there are no known defects and there are only minor things to be done. When the employment of the contractor is terminated under the contract, whether by the employer or by the contractor, the Works will never reach practical completion under that particular contract. The only way to complete the Works will be for the employer to enter into a new contract with another contractor. The new contract will not be for the Works as included in the original contract, but only for the balance of the Works.

When the new contract is completed, the architect will issue a certificate of practical completion for the Works included in the new contract. The certificate will be issued to the employer with a copy to the new contractor and it will refer only to the balance of the original Works. The original contractor will not receive a copy of this certificate, because it was not concerned in the new contract. The original contractor will not receive a copy of a certificate of practical completion of the Works in the original contract, because they were never completed.

It is surprising how often architects and contractors get confused about this.

115 Is the architect entitled to issue the practical completion certificate and an extension of time on the instruction of the client's solicitor?

This kind of question arises in many different guises. Architects are constantly concerned about what they should or could do so far are the contract is concerned. The golden rule is that the architect can only act if named in the contract. This architect must do all the things that the contract states the architect 'shall' do and may do all the things which the contract says the architect 'may' do.

Applying this simple rule to the question, we have to ask ourselves if there is anything in the contract which states that the architect may or even shall carry out functions described in the contract in the way instructed by the client's solicitor. The answer to that question is that no standard form of contract compels or even allows the architect to act in that way. Quite the reverse. Standard contracts invariably state that architects must exercise their own professional opinion when deciding matters such as practical completion or extensions of time; indeed when exercising any certification function. Even if the contract did not expressly so state, it would be implied. Put another way, for the architect to be compelled or allowed to comply with the client's solicitor's instructions in this way there would have to be an express term in the contract. The reason is obvious: if architects had to act in accordance with such instructions, there would be no need for the architect at all, certainly as contract administrator. Moreover, any certificates issued as a result of such instructions would carry no weight at all. They would simply be like statements issued by the employer under a design and build contract.[168]

Obviously, the situation is somewhat different if architects call for advice from their own legal advisers. But in such a case, architects are not being instructed by legal advisers; they are being advised. Like all advice, this is advice that architects are free to accept or reject. Therefore, it is still a matter for an architect's own opinion, albeit reached with the benefit of advice.

116 If the employer has agreed that the contractor can have another 4 weeks' extension of time, should the architect backdate the certificate of practical completion?

Under the standard form contracts where an architect is engaged, it is for the architect to decide on the amount of extension of time to which the contractor is entitled. There is no power under such contracts for the employer to agree extensions of time with the contractor. Of course, it is always open to the employer and the contractor to mutually agree to change the date for completion. However, that is not extending time, it is varying the terms of the original contract, something that only the parties to that contract can do.

Where the parties do agree a new date for completion, the agreement should be embodied in a formal document, which can be quite brief, to record their agreement. Alternatively it could be done by a carefully worded letter, one to the other, and signed by both parties. Such an agreement must only be concluded for a good reason, because under normal circumstances the architect will simply extend time as appropriate.

For the architect to backdate a certificate of practical completion as a way of putting into effect the parties' agreement is a recipe for chaos of the highest order. Backdating a practical completion certificate at all is wrong. It is a negligently issued certificate for which the architect has no possible excuse since it is a deliberate issuing of a certificate naming a date that the architect knows full well is 4 weeks earlier than the correct date. It not only certifies the wrong date, on which other contracts such as the beginning of a lease may depend, it also reduces the possible amount of liquidated damages the employer may recover, it starts the rectification period four weeks early with a consequent four weeks earlier finish, and it throws the onus of insurance on to the employer at a time when the contractor is still working on site. The date of practical completion determines the end of the period during which the contractor is entitled to be on site. If the contractor remains on site thereafter, it is a trespasser.

If the parties have agreed the change to the completion date, the backdating of the certificate of practical completion is superfluous, because the change is already agreed. The effect of the backdating is then to reduce the period from the contract completion date to practical completion by 8 weeks instead of 4.

There can be little excuse for the employer to become involved in changes to the completion date. These things usually happen because the contractor has made a direct approach to the employer. If the employer is well-briefed by the architect, the contractor's approach will be smartly repulsed and the contractor will be instructed to direct all queries to the architect.

If the employer is actually determined to become involved in running the project, the architect must seriously question whether it is possible to continue to administer the contract in accordance with its terms. The employer must be warned that the interference may amount to a repudiatory breach of the contract of engagement, with all that entails.

117 Can partial possession be used for the whole of the interior if the employer is anxious to move in?

SBC, IC and ICD all make provision for the employer to take partial possession of the Works in similar terms. Clause 2.33 of SBC provides that, if the contractor consents, the employer may take possession of any part or parts of the Works. Once partial possession occurs, the architect must supply a written statement to that effect, identifying the part taken into possession and practical completion is deemed to have taken place for that part. It should be noted that practical completion has not actually taken place, but the parties agree to treat it as though that were the case. It follows that the rectification period starts for that part and the contractor's duty to insure the part comes to an end. In addition, the amount of liquidated damages it is possible to levy in respect of the remaining part is to be reduced pro rata the value of the part taken into possession.

Employers should not be encouraged to take partial possession of virtually the whole building. The contract provides for completion of the Works by a stipulated date and if the contractor, through its own fault, fails to so complete, the employer has entered a sum of liquidated damages in the contract particulars to ensure compensation for the delay. Therefore, there really is no excuse for the employer wishing to take possession before completion. In practice, of course, employers do not always insert the full amount of liquidated damages for fear it will discourage contractors from tendering. Also, the employer may have miscalculated the liquidated damages amount, which may be insufficient to recompense for the loss suffered on account of the delay.

There can be little doubt that the sort of partial possession envisaged by the JCT draftsman is where one or two rooms or areas are taken into possession in advance of the main building. Therefore, the question whether the clause can be used to enable the employer to take into partial possession the whole of the building except for the external walls is difficult. On a strict reading of the contract, clause 2.33 of SBC does appear to allow such partial possession although it may be ill-advised for the employer to do so. That is because presumably there will be a substantial amount of internal work to be finished and defects to be rectified. If that were not the case, the architect would simply certify practical completion of the whole building.

It is important to understand that the architect's role in partial possession is passive. It is for the employer to request partial possession from the contractor and for the contractor to consent or otherwise. The architect's sole task is to issue the statement noted above. The employer could ask for partial possession of three rooms on the second floor, two on the first floor and a length of corridor on the ground floor. A difficulty with partial possession is that it may make working arrangements more difficult for the contractor. A contractor who is hindered in carrying out the Works is entitled to claim loss and/or expense under clause 4.23 of SBC.

If partial possession is taken of the whole of the interior, the architect, in issuing the written statement under clause 4.23, ought to include plans and sections of the building so that there is no doubt about the volume concerned. It must be made crystal clear whether the partial possession includes the internal face of the external walls and the ceiling of the topmost storey – common sense suggests that these surfaces must be included.

The architect's role is important in advising the employer of the dangers, principally of contractor's claims and loss of liquidated damages, of partial possession, whether it be of the whole interior or just a small part.

20 Termination

118 The contractor is running over time. The employer wishes to terminate, but the architect has over-certified.

An architect who certifies more than the amount properly due to the contractor is negligent and in breach of his or her contractual obligations to the employer by which the architect undertook to administer the contract. All the standard forms of contract permit the architect to certify only those amounts properly due. Because payment certificates are cumulative, the architect should be able to retrieve the situation in the next certificate. Obviously, that would not be the case if the employer terminated, the contractor went into liquidation, or the certificate in which the over-certification took place was the one issued after practical completion. Interim certificates are not intended to be precisely accurate. It has been well said that they are essentially a means by which the contractor is assured of some cash flow that roughly approximates to the work carried out.[169]

It is doubtful whether the employer would have grounds to terminate the contractor's employment simply because the contractor was running over time. Under JCT contracts, the architect would have to be satisfied that the contractor was failing to proceed regularly and diligently. However, for the sake of this question let us suppose that there are adequate grounds to terminate. The architect will be obliged to confirm that to the employer. Then the architect has no option but to inform the employer that there has been some over-certification and that it would be better to wait

169 *Sutcliffe v Chippendale and Edmondson* (1971) 18 BLR 149; *Sutcliffe v Thackrah* [1974] 1 All ER 319

until the work on site has caught up with the amount certified. How long that will be will depend on the degree of over-certification and the rate at which the contractor is currently working.

Of course, the employer can always go through the termination procedure and then reclaim the over-certified amount, together with any other balance due, after the Works have been completed by others. In doing so, however, the employer would have to challenge the architect's certificate, probably in adjudication. This would be additional cost which, if the adjudicator found in favour of the employer, the employer would assuredly try to recover from the architect.

In practice, instances of over-certification tend to be unusual and concerning relatively small amounts, usually the result of the valuation of defective work. It is the architect's responsibility to notify the quantity surveyor of all defects so that they will not be valued. In any event, certification is entirely the province and responsibility of the architect.[170] That is why architects should not simply accept valuations from quantity surveyors and blindly transfer the valuation figures to the certificate; they should ask for a simple breakdown of the valuation. It is not the architect's job to do the valuation again, but the architect should have enough information to be satisfied that the valuation looks about right.

119 Is there any time limit for the employer to terminate after the architect has issued a default notice?

Termination is such a draconian remedy that it has to be carried out correctly or it will not be valid. It is dangerous to get involved in an invalid termination situation. Any party involved in potential termination should always seek appropriate advice.

An example of an architect poorly handling a termination situation, but still muddling through, may serve as some comfort for those architects who habitually muddle through the intricacies of JCT contracts. The case was *Robin Ellis Ltd v Vinexsa International Ltd*.[171] It concerned the termination provisions of the

---

170   *R M Burden Ltd v Swansea Corporation* [1957] 3 All ER 243
171   [2003] BLR 373

IFC 84 form of contract, but the principles are applicable to the SBC, IC and ICD contracts.

The contract did not run smoothly and Ellis (the contractor) removed itself from site. Two days later, the architect sent a 14-day default notice on the basis that Ellis had, without reasonable cause, wholly suspended the carrying out of the Works. Ellis then went back to site and continued to work. Two weeks later, however, it left site again. This is where the architect's difficulty arose. Clause 7.2.3 states: 'If the Contractor repeats the specified default, then within a reasonable time after such repetition, the Employer may by notice to the Contractor terminate the employment of the Contractor under this Contract.' The current contracts use much the same words.

Unfortunately, the architect appears not to have read this provision. He issued another default notice giving a further 14 days to rectify the default. A few days later, the architect, realising the mistake, wrote and withdrew the notice. Not waiting until the expiry of 14 days from the second default notice, Vinexsa issued a termination notice to Ellis. The parties went to arbitration and then appealed to the court. Ellis alleged that the termination was not valid, because Vinexsa should have waited 14 days from the expiry of the second notice. Moreover, the architect did not have power to withdraw the second notice.

The court said that if a default had been made the subject of a default notice, a second notice could not be issued in respect of the same default. Therefore, the second notice was invalid. Therefore, its withdrawal was not in issue. Ellis then contended that the second default notice amounted to a representation that entitled Ellis to believe that it would have a further 14 days. The court accepted that, although the architect was wrong to have issued the second notice, having issued it, Vinexsa might be vexatious in immediately terminating. But the second default notice was invalid and it had been withdrawn; therefore Vinexsa was entitled to give notice of termination.

In some ways, the result was surprising; but the message is clear. After the default notice the employer has a specified time (10 days) in which to issue the notice of termination. If the employer opts not to do so or if the default is rectified and, in either case, the contractor repeats the same default at a later date, it is wrong to issue a fresh default notice and the employer must simply issue a notice of termination within a reasonable time. Therefore, the answer to the question is: within 10 days of the expiry of the default notice period or, on repetition, within a reasonable time thereafter. As usual, what is reasonable will depend on the circumstances.

120 The employer has terminated, but the contractor refuses to leave site, saying that it is entitled to stay until paid in full.

SBC, IC and DB (all in clause 8.7.1) and MW (in clause 6.7.1) provide that if the contractor's employment is terminated by the employer under the contractual provisions as the result of a default, the employer may employ other persons to complete the Works and that the employer and the other persons may 'enter upon and take possession of the site and the Works'.

Although the JCT contracts appear to cover the position fully (what can be clearer than saying that the employer and another contractor can take possession), it should be noted that it is qualified by reference to termination by the employer. Therefore, it appears to be open to the contractor to refuse to give up possession on the ground that the termination is invalid. If the contractor is able to do that, the progress of the Works could be held up for a long time while the parties argue the matter before adjudicator, arbitrator or judge. It would have been preferable for the contracts to have provided expressly that the contractor should give up possession whether the termination was disputed or not. The ACA Form of Building Agreement (ACA 3) has this provision and it can eliminate difficulties.

If the contractor does not dispute the validity, but merely states that it is entitled to remain on site until paid, such a contractor is doomed to failure. First, because, as has been seen, the contracts provide for possession to be taken by the employer. Second, because the contracts expressly state that the employer is not obliged to make any further payment until after the Works have been completed and any defects made good. The problem arises (but not under ACA 3) only if the contractor disputes the validity of the termination. If the contractor sits on site, the employer will probably apply to the court for an injunction to force the contractor to leave. An injunction will not usually be given if the court believes that damages will be a sufficient remedy. In the past, an injunction to remove a contractor after termination has been refused, because the contractor vehemently disputed the validity of the termination.[172] However, provided that the court is satisfied that all the procedural steps for termination have been properly carried out by the employer, it is likely that an injunction would be issued in a similar situation today.[173]

Where a contractor threatens to remain on site, it is usually to force some kind of agreement. If the employer stands firm, the wise contractor will usually leave rather than have an injunction served.

121 Termination took place due to the contractor's insolvency under SBC. The liquidator is insisting that full payment of any balance plus retention be immediately payable.

Termination due to the contractor's insolvency is covered by clause 8.5 of SBC. The employer may terminate at any time by written notice. The interesting point is that, under clause 8.5.3.1, as soon as the contractor becomes insolvent, even if no written notice has been given by the employer, the provisions of the contract that require any further payment or release of retention cease to apply.

That means that, although the employer may be slow in taking action to terminate the contractor's employment, the employer's duty to pay is at an end except as set out in clauses 8.7.4, 8.7.5 and 8.8. It should be noted, however, that there is no longer any provision for automatic termination.

These clauses stipulate that, within a reasonable time after the completion of the Works by another contractor and of the making good of defects, an account must be drawn up of whatever balance may be due from employer to contractor or vice versa as the case may be after taking into account all the costs of finishing off including any direct loss or damage caused to the employer by the termination. If the employer decides not to complete the Works using others, a statement of account must be drawn up and sent to the contractor after the expiry of 6 months from termination.

It is not unknown for liquidators, either directly or more usually through the services of specialist insolvency surveyors, to threaten employers with proceedings if payment of all money is not made immediately. This contract now makes clear that the

172 *London Borough of Hounslow v Twickenham Garden Developments* (1970) 7 BLR 81
173 *Tara Civil Engineering v Moorfield Developments* (1989) 46 BLR 72

normal payment provisions are at an end. Therefore, if a certificate has already been issued, the employer no longer has any obligation to pay the sum certified. If a certificate is due, the architect no longer has a duty to certify. The employer or the architect should respond to the liquidator or the surveyor, referring to the contract clause as the reason why no further payment will be made until the final statement of account is prepared.

In these circumstances, the employer will often have delayed payment of a certificate because it is suspected that the contractor is about to become insolvent. Indeed, the employer's failure to pay and the failure of others may be the very reason why the contractor eventually becomes insolvent. Therefore, at the date of insolvency, not only is a certified amount outstanding, but it may have been outstanding for a considerable time. Is the employer bound to pay such a sum on the basis that the employer cannot take advantage of its own breach of contract (failing to pay a certified amount on time)? The answer to the question seems to be that the employer is bound to pay monies that were outstanding at the date of the insolvency. The use of the words '*which require further payment*' appears to support that, because a payment that is overdue is not a '*further payment*', but rather a payment that ought to have been made already. The point is not beyond doubt and, in practice, the courts may be reluctant to order such payment when there is no realistic chance of it being recovered later if the cost of completion of the Works proves to be more than the cash still retained by the employer.

122 Is it true that, under SBC, if the employer fails to pay, the contractor can simply walk off site?

It is surprising how often a contractor will simply stop work because it is not being paid. Generally, it is small contractors, perhaps operating under MW, who are most prone to this kind of action, but it has certainly been done with much larger contracts. It is difficult not to have considerable sympathy with a contractor who says, quite reasonably, that there is no point in doing further work if it has not been paid for what has already been done. Nevertheless, the law does not generally allow the contractor simply to stop work if not paid.

Under SBC, the remedy available to the contractor if the employer fails to pay a certificate is to issue a 7-day written notice under clause 4.14 and then to suspend performance of all its obligations if payment is not made. Additionally or alternatively, the contractor may simply issue a default notice prior to termination of its employment under clause 8.1.1. In any event, the contractor is entitled to recover interest, usually at 8 per cent above Bank of England base rate.

What is often forgotten by an employer is that the contract does not remove the contractor's common law rights. Quite the reverse. Clause 8.3.1 expressly preserves both the employer's and the contractor's other rights and remedies. It is established that if payment is made so irregularly and inadequately that the day arrives when the contractor has no confidence that it will ever be paid again, it amounts to a repudiatory breach of contract which the contractor may accept, bring its obligations to an end and recover damages.[174]

More recently, in *C J Elvin Building Services Ltd v Peter and Alexa Noble*[175] it was held that the employer was in breach of contract, not only by refusing to pay sums as they became due, but also by threatening to make no further payment until the job was completed. The contractor was, therefore, justified in suspending the Works. Indeed, from the judgment, it was clear that the judge considered that the contractor could have accepted the employer's breach as repudiation.

Therefore, the next time a contractor walks off site for lack of payment, both architect and employer should think carefully about whether the contractor is simply exercising its common law rights.

---

174 *D R Bradley (Cable Jointing) Ltd v Jefco Mechanical Services* (1989) 6-CLD-07–19
175 (2003) CILL 1997

21 Disputes

123 An Adjudicator's Decision has just been received and it is clear that the points I made have been misunderstood and the adjudicator has got the facts wrong. Can enforcement be resisted?

The short answer to this is no, at least not on those grounds. When an adjudicator makes a decision, the parties must carry out that decision within whatever timescale the adjudicator has laid down. If a party fails to carry out the decision, the other party has the right to apply to the court to enforce the decision. The decision of an adjudicator, while not quite sacrosanct, is at least protected until the parties decide that one or other wants to have the dispute finally decided in arbitration or in legal proceedings.

Adjudicators are human like the rest of us. Sadly, some adjudicators seem to have a tenuous grasp of the law. Nonetheless, when an adjudicator has been nominated, or perhaps the parties have agreed the name between them, that person is the one entrusted to make a decision on the merits of the dispute. It may well be that the adjudicator misunderstands some of the points made, and some participants do not make themselves very clear. On the other hand, some participants are represented by experts who subject the adjudicator to a barrage of words. It is not unusual for an adjudicator to receive half a dozen or more lever arch files as the referral and a couple of similar-sized files as the response, to say nothing of a multitude of submissions on jurisdiction. It is little wonder if some of the more subtle points are overlooked in this scenario. Again, it must be said that some adjudicators are not good at handling clever points and generally try to come to a decision on what they believe is the overall justice of the case. Of course, this is not what an adjudicator is supposed to do. An adjudicator, just like an arbitrator or a judge, is charged with applying the law, not a personal gut feeling. Lord Denning may have been famous for that very thing, but he was an exception and in any event he could never have been accused of not knowing the law.

Adjudication was never intended for this kind of detailed argument. It was originally devised as a method of getting a quick result to problems that regularly bedevil the construction industry. The principle is that, if the adjudicator answers the right question in the wrong way, the decision will be upheld by the courts, but if the adjudicator answers the wrong question in the correct way, the decision will be a nullity. Another way to put it is that the adjudicator can answer only the question asked in the notice of intention to seek adjudication. Neither party can unilaterally introduce new questions; there can be no counterclaims. The only exception is if the adjudicator has to answer a question that has not been asked in order to answer the question asked.[176] Even if the adjudicator has misunderstood the facts or simply got them wrong, it is not enough to resist enforcement of what might well be a flawed decision. This was established by two early cases.[177]

Basically, it is a policy decision. Adjudication is a quick method of settling disputes, but a comparatively coarse remedy. With this kind of remedy, the parties have to accept that there will be rough justice and, occasionally, even bad errors. If the money at stake is sufficiently large, no doubt the parties will seek a solution in a forum that can give proper time and consideration to the arguments.

The main acceptable grounds for resisting enforcement of a decision tend to concern whether the adjudicator had the jurisdiction to make the decision. If it can be shown that there was no jurisdiction, the enforcement can be resisted, because there is no decision. Lack of jurisdiction is often due to a failure to answer the question asked, so that the adjudicator has no jurisdiction for the decision about an unasked question. It may also be because it can be shown that there was no dispute in being at the time the notice of intention

176 *Karl Construction (Scotland) Ltd v Sweeney Civil Engineering (Scotland) Ltd* [2001] SCLR 95

177 *Macob Engineering v Morrison Construction* [1999] BLR 93; *Bouygues (UK) Ltd v Dahl-Jensen (UK) Ltd (in Liquidation)* [2000] BLR 49

to seek adjudication was issued. Successful challenges have also been made on the basis that the adjudicator was in breach of the rules of natural justice. This can be quite complicated in practice, but in essence it refers to the need for each party to have an opportunity to put its case. For example, the adjudicator must not discuss the dispute with one party in the absence of the other. Obviously, a decision will also be thrown out if it can be shown that an adjudicator was biased in favour of the other party.

124 An Adjudicator has been appointed whom the employer has not agreed. What can the employer do about it?

The adjudicator can be appointed in various ways. The JCT Standard Building Contract 2005 (SBC) has provision in clause 9.2.1 and in the contract particulars for the adjudicator to be named, i.e. the parties have decided on the name before entering into the contract. There is also provision for the parties to simply apply to one of several bodies that maintain panels of adjudicators and which will undertake (for a modest fee) to nominate an adjudicator who has passed the criteria set by that particular body. The RIBA, RICS and the CIArb are typical nominating bodies. Of course, there is nothing to stop the parties simply agreeing the name of an adjudicator when the dispute arises. If the parties can agree a name, that is by far the best way, because the adjudicator is then a person whom both parties trust to make the right decision.

Putting the name in the contract in advance appears to have the same effect, but it does have some serious disadvantages. An adjudicator chosen in advance, perhaps several months or even years before required, may not be available when required due to holidays, illness, retirement or even death. The adjudicator may be admirable in very many respects, but not suited to the particular problem that arises. For example, the parties may choose Mr X, who is an architect, but the eventual dispute may concern structural steel or electrical services. The danger of simply applying to a nominating body is that the parties are stuck with whoever is nominated, unless they both object and agree to ask the adjudicator to step down in favour of another of their choosing.

Therefore, the answer to this question depends on what the contract says or, if there is no standard contract, what the Scheme for Construction Contracts (England and Wales) Regulations 1998 says.

It is essential that the adjudicator is appointed in strict conformity with any procedure that is laid down. If the relevant procedure has been complied with, there is nothing the employer can do about it. The adjudicator must be accepted. Obviously, if there is a good reason to object, for example if the adjudicator has a link with the other party, the appointment can be challenged, and if the adjudicator does not step down the employer can agree to proceed with the adjudicator without prejudice to the contention that the adjudicator lacks jurisdiction on the grounds of actual or apparent bias or an apparent breach of the rules of natural justice. Then that position can be tested in the courts if the adjudicator's decision is adverse to the employer.

If the proper procedure for the appointment of the adjudicator has not been observed, the employer is on firm ground to challenge and the adjudicator should resign as soon as it is established that the appointment is flawed.

125 Is the architect obliged to respond to the referral on behalf of the employer if so requested?

This is not as unusual a question as may seem at first sight. Most employers have no idea about adjudication. When they receive a notice of intention to seek adjudication from a contractor, they immediately send it to the architect who has been handling the project for them. No doubt, if there is a project manager, the employer will direct the notice there instead. But, whether project manager or architect, is there a duty to deal with the adjudication? Clearly the answer is 'No'.

So far as the architect is concerned, his or her duties will probably include taking the brief, preparing designs, submitting for planning and building control, preparing detailed construction drawings and administering the contract with all that entails, including dealing with contractor's claims (usually at an additional fee). It is understood that the architect is skilled and experienced at doing all these things. But the architect has neither the skills nor training to run an adjudication. The architect is no more capable of doing that than running an arbitration or dealing with court procedures. None of these things will be among the duties the architect has agreed to carry out. Importantly, the architect will have no professional indemnity insurance to cover such work.

Some architects may have special expertise in dealing with adjudications, just as some architects have expertise at acting as expert witnesses or designing particular types of buildings. In that case, there is no reason why such architects should not run the adjudication on behalf of the employer (again, for an additional fee). They should always remember that the adjudication is between the contractor and the employer, not between the contractor and the architect, even though the contractor may be invoking adjudication to question an architect's decision.

Although architects have no duty to run the adjudication for the employer, they clearly do have a duty to give the employer basic initial advice when the notice is brought to their attention. Architects should be capable of doing that, and it will usually consist of advising their clients to seek appropriate legal advice without delay in view of the alarmingly short time scale involved.

Architects, together with all the other consultants, certainly have duties to assist the employer's legal advisers by providing copy correspondence, drawings and, if appropriate, explanations about various aspects of the project. An appropriate fee will be chargeable, probably on a time basis. However, where the information required by the employer's legal advisers is more than merely factual (correspondence and drawings) consultants should tread warily. If the adjudicator's decision goes against the employer, the employer's legal advisers may start to look very carefully at the opinions and reports provided by consultants to see if the employer has a means of redress in that direction. It is not going too far to suggest that an architect or other consultant who is asked to assist in this way should consult their own advisers with a view to protecting their positions in the future.

An architect recently asked whether or not there was a duty to assist the employer in an adjudication that had been commenced by the contractor some time after the employer had decided to do without the services of the architect half-way through the project. The employer's legal advisers were pressing the architect to attend meetings and even to decide matters of extension of time and loss and/or expense. In such situations, the architect has no duty other than providing the employer with copies of documents on payment of reasonable copying expenses. By dispensing with the architect's services, the employer was no longer relying on the architect's skill and care. If the architect was asked to provide a witness statement as to facts, it is arguable that the architect could refuse. An architect asked to give evidence in arbitration or litigation could face a subpoena to do so. An employer who had sacked the architect would probably have little to gain by trying to force the architect to give evidence in such cases.

Table of cases

Ex parte Sir W Harte Dyke. In re Morrish (1882) 22 ChD 410 *161*
Fairweather (H) & Co Ltd v London Borough of Wandsworth (1987) 39 BLR 106 *174*
Finnegan (J J) Ltd v Ford Seller Morris Developments Ltd (No.1) (1991) 25 Con LR 89 *53, 143, 185*
Formica Ltd v ECGD [1995] 1 Lloyd's Rep 692 *65*
Freeman & Son v Hensler (1900) 64 JP 260 *103*
Gallagher v Hirsch (1899) NY 45 *119*
George Fischer Holdings Ltd v Multi Design Consultants Ltd and Davis Langdon & Everest (1998) 61 Con LR 85 *81*
Gilbert Ash (Northern) v Modern Engineering (Bristol) [1973] 3 All ER 195 *29*
Gleeson (M J) (Contractors) Ltd v Hillingdon Borough Council (1970) 215 EG 165 *177*
Glenlion Construction Ltd v The Guinness Trust (1987) 39 BLR 89 *40*
Gray (Special Trustees of the London Hospital) v T P Bennett & Son (1987) 43 BLR 63 *83*
Greater London Council v Cleveland Bridge & Engineering Ltd (1986) 8 Con LR 30 *85*
Hadley v Baxendale (1854) 9 Ex 341 *93*
Hall & Tawse South Ltd v Ivory Gate Ltd (1997) 62 Con LR 117 *6*
Hampshire County Council v Stanley Hugh Leach Ltd (1991) 8-CLD-07–12 *41, 50*
Harbutt's Plasticine Ltd v Wayne Tank and Pump Co Ltd [1970] 1 All ER 225 *162*
Haulfryn Estate Co Ltd v Leonard J Multon & Partners & Frontwide Ltd, 4 April 1990 unreported *99*
Hedley Byrne & Co Ltd v Heller & Partners Ltd [1963] 2 All ER 575 *57, 64*
Henderson v Merritt Syndicates [1995] 2 AC 145, (1994) 69 BLR 26 *59*
Henry Boot v Alstom Combined Cycles [1999] BLR 123 *115*
Higgins (W) Ltd v Northampton Corporation [1927] 1 Ch. 128 *16*
Holland Hannen & Cubitts (Northern) Ltd v Welsh Health Technical Organisation (1981) 18 BLR 80 *35*
How Engineering Services Ltd v Lindner Ceilings Partitions plc, 17 May 1995 unreported *45*
Impresa Castelli SpA v Cola Holdings Ltd (2002) 87 Con LR 123 *166*
In re Wheatcroft (1877) 6 Ch 97 *65*
Inserco v Honeywell, 19 April 1996 unreported *117*
Investors in Industry Commercial Properties Ltd v South Bedfordshire District Council (1986) 5 Con LR 1 *97*

Index